Communications in Computer and Information Science 1725

More information about this series at https://link.springer.com/bookseries/7899

Henry Han · Erich Baker (Eds.)

The Recent Advances in Transdisciplinary Data Science

First Southwest Data Science Conference, SDSC 2022
Waco, TX, USA, March 25–26, 2022
Revised Selected Papers

 Springer

Editors
Henry Han (iD)
Baylor University
Waco, TX, USA

Erich Baker
Baylor University
Waco, TX, USA

ISSN 1865-0929 ISSN 1865-0937 (electronic)
Communications in Computer and Information Science
ISBN 978-3-031-23386-9 ISBN 978-3-031-23387-6 (eBook)
https://doi.org/10.1007/978-3-031-23387-6

This Springer imprint is published by the registered company Springer Nature Switzerland AG
The registered company address is: Gewerbestrasse 11, 6330 Cham, Switzerland

Preface

Data science is influencing more and more aspects of business, health, science, and engineering. Different data-driven techniques are being developed to customize different types of data from different fields. For example, AI is highly involved in trading and investment in finance. Machine learning applications are found not only in healthcare and biomedical data analytics but also in heliophysics, fluid dynamics, and chemical engineering. These data science techniques provide a powerful way to exploit large or even massive amounts of data in comparison to the traditional model-driven methods, which generally have theoretical assumptions that may not match the data reality.

Although this transdisciplinary field may still need time to become an independent field, data from different fields are driving unprecedented expansion of the data science domain as its scope becomes broader and deeper. Investigators from different disciplines are generating new knowledge and techniques to extend data science from more specific applications. For example, researchers in astrophysics and business may both investigate the applications of Convolutional Neural Networks (CNN) in their fields, but the former may want to conduct glitch detection for gravity wave data whilst the latter seek to design an efficient trading machine by exploiting the powerful learning capabilities of the deep learning model. The different types of data may generate different CNN techniques for the corresponding problem solving and thereby enrich AI and machine learning.

The future of data science lies in the generation of different subdomains according to different data, with AI and machine learning serving the core of the general data science domain. The corresponding methods in each subdomain will be further tailored and optimized to accommodate different subdomain data. It is expected that machine learning and deep learning methods will be developed and defined according to specific types of data. For example, a transformer model will be redefined or polished differently under financial data, biomedical data, and social science data. Furthermore, the data science subdomains, such as financial data science, health data science, social data science, cyber data science, or even data engineering, etc., will interact with each other through explainable AI to enhance the interpretability of the results and techniques of different data science subdomains.

This volume (CCIS 1725) aims to report recent advances in business data science, health data science, applied data science, artificial intelligence, and data engineering. It collates the work of data science researchers from business, bioinformatics, computer science, health, physics, and engineering fields. We hope that tthe unique contents, including novel techniques in the different data science domains, will serve as a good guide for data scientists and practitioners in related fields.

This volume contains the best papers from the Southwest Data Science Conference (SDSC) 2022, which was organized by the Data Science group at Baylor University, Waco, Texas, USA, and took place during March 25–26, 2022. SDSC 2022 attracted a total of 72 submissions, consisting of 49 full papers and 23 long abstracts, which underwent rigorous single-blind peer review. Each paper was reviewed by at least three

Program Committee members and finally XXX papers were accepted for presentation at the conference and inclusion in this proceedings.

Henry Han
Erich J. Baker

Organization

General Chair

Henry Han — Baylor University, USA

Program Chair

Erich J. Baker — Baylor University, USA

Steering Committee

Henry Han — Baylor University, USA

Erich J. Baker — Department of Computer Science, Baylor University, USA

James Stamey — Department of Statistics, Baylor University, USA

Jonathan Trower — Hankamer Business School, Baylor University, USA

Greg Hamerly — Department of Computer Science, Baylor University, USA

Greg Speegle — Department of Computer Science, Baylor University, USA

Liang Wang — Department of Public Health, Baylor University, USA

Keith Schubert — Electrical and Computer Engineering, Baylor University, USA

Program Committee

Mary Lauren Benton	Baylor University, USA
Liang Dong	Baylor University, USA
Jeff Forest	Slippery Rock University, USA
Michael Gallaugher	Baylor University, USA
Greg Hamerly	Baylor University, USA
Shaun Hutton	Baylor University, USA
Chan Gu	Ball State University, USA
Haiquan Li	University of Arizona, USA
Zhen Li	Texas Woman's University, USA
Bill Poucher	Baylor University, USA
David Peng	Lehigh University, USA

Eunjee Song	Baylor University, USA
Greg Speegle	Baylor University, USA
Yang Sun	California Northstate University, USA
Ye Tian	Case Western Reserve University, USA
Stellar Tao	UT Health Center, USA
Liang Wang	Baylor University USA
Jeff Zhang	California State University, Northbridge, USA

Keynote Speaker

Mark Ferguson	University of South Carolina, USA

Additional Reviewers

Yaodong Chen	Northwest University, China
Heming Huang	Qinghai Normal University, China
Chun Li	Hannai Normal University, China
David Li	Institute of Plant Physiology and Ecology, China
Wenbin Liu	Guangzhou University, China
Joe Ji	Columbia University, USA
Jennifer Ren	Baylor University, USA
Juan Wang	Qufu Normal University, China
Jay Wang	Monmouth University, USA
Yi Wu	Georgia Institute of Technology, USA
Juanying Xie	Shaanxi Normal University, China
Liang Zhao	University of Michigan, Ann Arbor, USA

The Convergence of BI and AI: With Applications for the Better Management of Higher Education (Keynote)

Mark Ferguson

Associate Dean for Academics and Research, The Dewey H. Johnson Professor of Management Science Darla Moore School of Business, University of South Carolina, Columbia, SC 29028, USA
mark.ferguson@moore.sc.edu

Abstract. The terms Artificial Intelligence (AI) and Machine Learning (ML) have been around for some time but has obtained a much broader exposure during the last five years. This increase in exposure seems to be coming at the expense of Business Intelligence (BI), as more university classes and company projects are being rebranded from BI to AI. The producers of BI tools and solutions most frequently refer to themselves at Business Analysts, while the producers of AI tools and solutions refer to themselves as Data Scientist. I'll first present a case that the biggest opportunities for most organizations is in the convergence and tighter integration of these two areas. As an example application of the potential of this convergence, I'll discuss an industry that should be familiar to most of the audience, higher education. Despite employing arguably some of the best talent available in both the AI and BI space, the administration of most colleges and universities lags far behind in the use of these powerful tools as compared to similar sized organizations in other industries. I'll share some examples of how the tools are being employed at my own university and postulate on some of the factors that inhibit wider adoption.

Keywords: AI · Business Intelligence · Higher education management

Dr. Mark Ferguson is the Senior Associate Dean for Academics and Research and the Dewey H. Johnson Professor of Management Science at the Darla Moore School of Business at the University of South Carolina. He received his Ph.D. in Business Administration with a concentration in operations management from Duke University in 2001. He holds a B.S. in Mechanical Engineering from Virginia Tech and an M.S. in Industrial Engineering from Georgia Tech.

Contents

Business and Social Data Science

Forecasting Stock Excess Returns with SEC 8-K Filings

Henry Han[1(✉)], Yi Wu[2], Jie Ren[1], and Li Diane[3]

[1] Department of Computer Science, Baylor University, Waco, TX 76798, USA
Henry_Han@baylor.edu
[2] Georgia Institute of Technology, Atlanta, GA 30332, USA
[3] School of Business & Technology, University of Maryland, Eastern shore, MD 21853, USA

Abstract. The stock excess return forecast with SEC 8-K filings via machine learning presents a challenge in business and AI. In this study, we model it as an imbalanced learning problem and propose an SVM forecast with tuned Gaussian kernels that demonstrate better performance in comparison with peers. It shows that the TF-IDF vectorization has advantages over the BERT vectorization in the forecast. Unlike general assumptions, we find that dimension reduction generally lowers forecasting effectiveness compared to using the original data. Moreover, inappropriate dimension reduction may increase the overfitting risk in the forecast or cause the machine learning model to lose its learning capabilities. We find that resampling techniques cannot enhance forecasting effectiveness. In addition, we propose a novel dimension reduction stacking method to retrieve both global and local data characteristics for vectorized data that outperforms other peer methods in forecasting and decreases learning complexities. The algorithms and techniques proposed in this work can help stakeholders optimize their investment decisions by exploiting the 8-K filings besides shedding light on AI innovations in accounting and finance.

Keywords: Form 8-K · Excess return · Dimension reduction stacking · NLP

1 Introduction

With the surge of AI and social data science, more and more Securities Exchange Commission (SEC) filing data and relevant information are employed in AI-based decision-making in accounting and finance [1, 2]. The SEC filing data mostly refers to those periodic (SEC) forms (e.g., 8-K, 10-K, 13-D), financial statements, and other required disclosures submitted by public companies. They disclose different important corporate actions and activities, the change of ownerships, or other significant information to investors. Since almost all SEC filings are textual data containing important sentimental information about companies, different natural language processing (NLP) and machine learning (ML) techniques are employed to predict the security tendency of the firms, identify possible artifacts in corporate management, or even provide information for strategic decision making [1–3]. For example, since an important acquisition reported in the SEC 8-K report may signal future corporate stock prices or other corporate actions, machine learning or deep learning method are used to query future stock price movements or predict the coming firm events or activities [3–5].

H. Han and E. Baker (Eds.): SDSC 2022, CCIS 1725, pp. 3–18, 2022.
https://doi.org/10.1007/978-3-031-23387-6_1

Form 8-K, which is known as the "current report", provides timely notification to shareholders about the occurrence of significant events. It is generally not filed at a specific time interval. When certain types of important events (e.g., acquisition) that should be aware by shareholders happen, an 8-K report should be filed in four business days. Unlike other information that can be modified or even interpreted by a third party, the 8-K information is valuable and reliable because it is direct communication between the companies and investors. Also compared to its similar peers such as Forms 10-K and 10-Q, Form 8-K is more concise, readable, and informative, especially because the 10-K/Q filings cannot timely notify investors about the important corporate events due to their long release time intervals.

Many recent efforts were invested in accounting and finance fields by employing NLP and AI-based textual analytics to study the SEC filings to decipher market trends, discover future corporate events, or predict security returns. Ke *et al.* introduced novel text-mining methods to extract sentiment information from news articles to predict stock returns. Loughran revealed meaningful liability signals by analyzing 10-K reports [4]. Zhai applied deep learning methods to predict firm future event sequences using the 8-K reports [5]. Kogan *et al.* predicted stock return volatility from 10-K forms [6]. Lee *et al.* used the 8-K filings to predict stock returns [7]. Zhao categorized different 8-K reports into different categories to seek signals for stock returns [8]. Furthermore, Engelberg found that textual information has better long-term predictability for asset returns [9]. Li employed a Naïve Bayes learning approach to analyze corporate filings [10]. Aydogdu *et al.* used long short-term memory (LSTM) deep learning models to analyze SEC 13D filings [11].

The studies bring new insights into finance and accounting fields besides extending AI application domains. However, they have the following weakness and limitations. First, the results from the proposed methods are not good enough for real business practice because of the high nonlinearity of data and possible artifacts of the methods. Few improvements were shown that stock movement prediction based on the textual data was better than using classic quantitative features [7]. This is especially true for those stock return forecasting or relevant studies using the 8-K filings [1, 3].

Second, it remains unknown how to handle the high dimensional vectorized data of the SEC filings well for the sake of the following ML prediction. The vectorized data obtained from the text vectorization process, which is a procedure to convert the original textual data (e.g., 8-K reports) into its numerical representation, can demonstrate very high dimensionality. For example, an 8-K report textual dataset with 10000 sentences can be converted to a corresponding numerical matrix with 10000 observations across 20000+ features after vectorization. In addition, the high-dimensional vectorized data may contain different noise from the original text and vectorization procedures. The traditional viewpoint believes dimension reduction is necessary or even a must before any serious downstream analysis such as machine learning.

However, it remains unknown how to conduct meaningful dimension reduction to gain effective feature extraction to balance the tradeoff between preserving useful information and removing redundancy besides de-noising. The widely used latent semantic indexing (LSI), an SVD-based dimensional reduction algorithm, or its variants (e.g., PCA) may not be able to represent the original data well in the low-dimensional space.

This is because they are holistic methods only good at capturing global data characteristics rather than local data characteristics [12]. The global and local data characteristics reflect those holistic (global) and local data behaviors respectively. To achieve high-performance learning, it is needed to retrieve both global and local data characteristics from the vectorized data. However, it remains unclear how to extract both global and local data characteristics in dimension reduction in NLP analysis not to mention for the SEC filing data.

Third, it remains unknown which text vectorization models match input textual data (e.g., 8-K data) well though different models are available. They include Bidirectional Encoder Representations from Transformers (BERT), term frequency-inverse document frequency (TF-IDF), and Bag-of-words (BoW), etc. [13, 14]. Since most studies only select one vectorization model to process textual data, the following ML results may not be optimal in business problem solving. Therefore, more comparisons and analyses are needed to examine the match between the ML methods and the vectorization models for the sake of decision making.

In this study, we address the challenges by forecasting the excess returns of SP500 companies using their 8-K filings. The excess return evaluates how a stock or portfolio perform in comparison to the overall market average return. It is generally represented as an Alpha metric computed by comparing with a benchmark index (e.g., SP500 index). The alpha value provides traders and portfolio managers a robust index to select securities or diversify portfolios in investment. Since the 8-K reports include the significant events of a company that would impact its stock performance, it will be interesting to predict the excess returns by exploiting the semantics of the 8-K filings. Therefore, predicting the excess return using the 8-K data not only explores more hidden relationships between the important corporate events and company stock performance, but also provides valuable insights for trading, portfolio management, and investment decision-making [15].

The stock excess return forecasting is modeled as an imbalanced multiclass classification problem in this study. We employ the data used in Ke *et al.*'s study which includes the 8-K filings from the SP500 companies during the period 2015–2019 [3]. The daily excess return of each stock, i.e., daily alpha values, is used as a target variable in forecasting. The stock excess return prediction is rendered as an imbalanced multi-classification problem by discretizing the target variables as three classes, because the number of observations in the majority class counts >50% among all observations. Unlike other classification problems in textual analytics with roughly balanced label distributions, the imbalanced learning problem itself not only challenges the excess return forecasting but also ML itself for its imbalanced nature.

We propose a multiclass support vector machines (SVM) model for the excess return forecasting for its good efficiency and reproducibility [16, 17]. We compare it with different peers under various dimension reduction and vectorization methods. Unlike the general assumptions, we find that dimension reduction may not contribute to enhancing the excess return prediction. They generally bring slightly lower or at most equivalent forecasting performance though they lower the complexity of the SVM learning. Moreover, inappropriate dimension reduction methods (e.g., PCA + tSNE) may generate an imbalanced point in the forecast. The imbalanced point is a special overfitting state in which the ML model misclassifies all minority samples as the majority type. Furthermore, we find more complicated deep learning models (e.g., transformer) are more likely to generate the imbalanced point than those with simple learning topologies (e.g., k-NN), besides resampling techniques may not be able to enhance the forecast performance.

In addition, we propose a novel dimension reduction stacking technique in this study to capture both global and local data characteristics of the vectorized data. The data under the new dimension reduction technique achieves an equivalent performance in comparison with using the original data besides decreasing the learning complexity of the stock excess return forecast. To the best of our knowledge, it is the first time that such a technique has been proposed and will bring positive impacts on meaningful NLP dimension reduction. Moreover, in contrast to the general viewpoint that BERT is more suitable for large-scale text data vectorization than TF-IDF for its more semantics-oriented vectorization, we find that TF-IDF leads over BERT in the stock excess return forecast [18].

This paper is structured as follows. Section 2 introduces the 8-K data preprocessing along with text vectorization. It also models the excess return forecast as an imbalanced learning problem and introduces imbalanced point generation and an explainable metric: d-index for the stock excess return forecast evaluation. Section 3 introduces the dimension reduction stacking technique for the global and local data characteristics retrieval for the vectorized data. Section 4 demonstrates the advantages of the proposed SVM model by comparing it with different peers under different dimension reduction methods and vectorizations. It also investigates how resampling techniques impact the stock excess return forecast. Section 5 discusses the improvements of the proposed techniques and concludes this study.

2   Data Preprocessing, Text Vectorization, and Imbalanced Learning Forecast

The goal of this study is to predict daily excess returns (alpha values) for SP 500 stocks with the 8-K data. We employ the 8-K filings from all the SP500 companies during the period 2015–2019 used in Ke *et al.*'s study [3]. The original textual data includes a total of 119762 sentences from collected 8-K documents [18]. The stock daily excess return, i.e., daily alpha, is used as a target variable in learning. We have valid 17648 daily stock excess returns by removing all missing data in preprocessing.

We model the stock excess return prediction with the 8-K data as an imbalanced multi-classification problem by discretizing the daily alpha values into three classes. The three classes demonstrate imbalanced distributions: the majority class counts > 50% of the total observations. This is because we partition the daily alpha values into three classes: class 0: alpha < -0.01, class 1: alpha in $(-0.01, 0.01)$, and class 2: alpha > 0.01. We simply call them 'buy', 'hold', and 'sell' though they do not represent real trading recommendations. The 'hold' class is the majority class with 9109 samples counting 51.61% of all observations. The 'buy' and 'sell' classes have 4138 and 4401 samples counting 23.45% and 24.94% of all observations, respectively. The preprocessed 8-K data is further partitioned as training data with 13655 samples and test data with

3993 observations. The three class distribution ratios among the test data are 25.44%, 48.01%, and 26.55%.

The stock excess return forecast using the 8-K data is challenging ML for its imbalanced nature. Imbalanced learning would bring biases to the classification accuracy metric so that it cannot reflect the true forecasting correctly [19]. This is because the forecast can be 'hijacked' by the majority class in learning, i.e., the ML model misclassifies most or even all minority observations as the majority type. In other words, the whole forecast is 'overfitted' to the majority type. As a result, the ML model loses its learning capability, and the accuracy will approach the majority ratio of the test data. If the ratio is high enough, the ML model will achieve deceptively decent performance even if it loses its learning capability. We say the ML model will generate an imbalanced point in this situation.

To avoid the accuracy bias from imbalanced learning, we employ an interpretable ML assessment metric diagnostic index (d-index) to evaluate the stock excess return forecast [19]. The d-index falls in (0,2] and a large d-index indicates better ML performance. The imbalanced point and d-index are described as follows. More detailed information can be found in the authors' recent work [19].

Imbalanced Point. Given training data $X_r = \{x_i, y_i\}_{i=1}^m$, in which $y_i \in \{1, 2, ..k\}$ is the label of the observation x_i, under an ML model Θ, we assume the class k is the majority type class and the majority ratio is defined as $\gamma = \frac{|\{x_i:y_i=k\}|}{\sum_{i=1}^k |\{x_j:y_j=i\}|} \gg \frac{|\{x_j:y_j=i\}|}{\sum_{i=1}^k |\{x_j:y_j=i\}|}, i \neq k$. Then the ML model is said to reach an imbalanced point, provided $\hat{f}(x|\Theta, X_r) = k$ for $\forall x$ whose label is unknown, in which $\hat{f}(x|\Theta, X_r)$ is the prediction function built under the model Θ with the training data X_r. The accuracy of the ML model will be the majority ratio γ at the imbalanced point if the training and test data share the same majority ratio. Otherwise, the accuracy will be the majority ratio of the test data.

d-index. Given an implicit prediction function $\hat{f}(x|\Theta) : x \rightarrow \{-1, +1\}$ constructed from training data $X_r = \{x_i, y_i\}_{i=1}^m$ under an ML model Θ, where each sample $x_i \in R^p$ and its label $y_i \in \{-1, 1\}, i = 1, 2, \cdots m$, d-index evaluates the effectiveness of $\hat{f}(x|\Theta)$ to predict the labels of test data $X_s = \{x_j', y_j'\}_j^l$, where x_j' is a test sample and its label $y_j' \in \{-1, 1\}$. The d-index is defined as:

$$d = \log_2(1 + a) + \log_2\left(1 + \frac{s+p}{2}\right) \tag{1}$$

where a, s, and p represent the corresponding accuracy, sensitivity, and specificity in diagnosing test data respectively. The d-index under multiclass classification can be extended from the binary version definition. More details about the d-index can be found in [19].

The d-index provides a good metric to compare the performance of different ML models for the sake of more robust model selection, especially under imbalanced learning. It is more explainable and comprehensive than other measures used for imbalanced learning evaluation such as MCC (Matthews correlation coefficient) and weighted F1 score [19]. The following theorem states the special d-index value at the imbalanced point. The detailed proof is omitted to conserve space.

Theorem 1. Imbalanced Point Theorem. The imbalanced point under an ML model Θ has an accuracy of γ and a d-index of $\log_2\left(\frac{3(1+\gamma)}{2}\right)$, where γ is the majority ratio of the test data.

In practice, we also count an approximately imbalanced point (AIP), in which the d-index of the ML model approximates $\log_2\left(\frac{3(1+\gamma)}{2}\right)$, as an imbalanced point. This is mainly because it may misclassify only a small portion of majority observations or not classify a few minority observations into the majority type. The AIP also indicates that the ML model loses its learning capabilities.

2.1 Text Vectorization

Text vectorization is an important feature extraction procedure to translate input text documents to their numerical matrix representations. There are several text vectorization methods such as BERT, TF-IDF, BoW, etc. to convert input 8-K text data into its corresponding numerical representations. Although it is generally unknown which text vectorization models will vectorize the 8-K data better for the following forecast, we have the following reasons to choose BERT and TF-IDF rather than BoW [14].

BERT is a bidirectional encoder that captures more context semantics in text vectorization. BERT employs the bidirectional training of Transformer neural networks in vectorization and the other peers only conduct one direction training or scan. Therefore, it captures more context information and produces more informative feature extractions. BERT is good at vectorizing large-scale text data (e.g., 8-K) because it is powered by a deep learning transformer model. The fine-tunning process allows BERT to learn more semantics from text data.

Both BERT and TF-IDF can avoid the negative impacts of the high-frequency words from the 8-K reports. The 8-K documents generally have similar words with very high frequencies to appear due to the filing syntax formats, most of which can be less significant or even meaningless. TF-IDF calculates the inverse document frequency to emphasize the significance of each word to avoid this issue. The semantics-oriented BERT avoids this by the bidirectional scan in vectorization. On the other hand, BoW relies on counting the frequency of each word in the document to build feature vectors to represent the original text data. Thus, the high-frequency words would bring noise in the BoW vectorization and impact the following forecasting negatively. Furthermore, since grammar information is completely lost in BoW, it can be hard to capture those features representing context semantics.

Figure 1 compares the visualizations of the BERT, TF-IDF, and BoW vectorized data with PCA. They can be viewed as the vectorization signatures of the three methods. It shows that all data points under BoW are jammed in a small region. This suggests that most of the data points of the vectorized data have close or similar variances. As such, it is almost impossible to distinguish different types of stock excess returns from the BoW data. This indicates that BoW may not be a suitable vectorization model for the 8-K data. On the other hand, both TF-IDF and BERT show a good spread of the data points and three types of returns are detectable though the majority class ('hold') still distinguishes itself. It is reasonable to expect that they will impact following forecasting more positively than BoW.

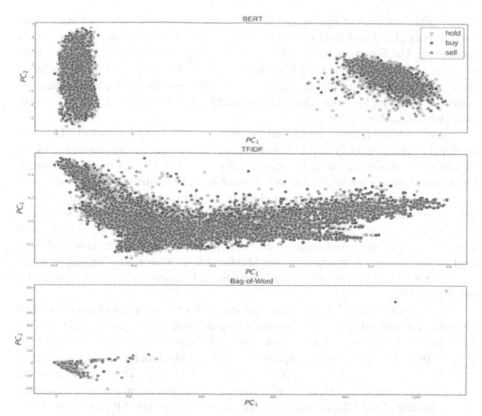

Fig. 1. The comparisons of the PCA visualizations of the BERT, TF-IDF, and BoW vectorization data, in which the daily stock excess returns are partitioned into three groups: 'buy', 'hold', and 'sell'. The BERT and TF-IDF vectorizations show a reasonable spread of three types of excess returns, but the BoW vectorization shows all data points are condensed in a small region.

BERT produces a relatively lower number of features than TF-IDF and BoW for its semantics-oriented vectorization. The BERT vectorized data has only 768 features, but the TF-IDF vectorized data has 27,147 features that is much larger than the number of observations 17,648.

3 Dimension Reduction Stacking

Since the vectorized data of the 8-K textual data is characterized by their high dimensionality and nonlinearity, dimension reduction is assumed as an essential to remove redundancy, decrease noise, and reduce forecasting complexities. The widely used dimension reduction methods: LSI, PCA, LDA, and manifold learning methods (e.g., t-SNE) are generally employed to accomplish it [20]. However, it remains unknown how to extract both global and local data characteristics from the dimension reduction procedures because they are either holistic or local dimension reduction methods. High-performance forecasting needs both global and local data characteristics to achieve good prediction.

We address the challenge by proposing a novel dimension reduction stacking to capture both global and local data characteristics for the sake of stock excess return forecasting. The dimension reduction stacking is to map input data $X = \{x_i\}_{i=1}^n$, $x_i \in R^p$ to its low-dimensional embedding $Y = \{e_i\}_{i=1}^n$, $e_i \in R^l$, $l \ll p$ by stacking two different dimension reduction algorithms: $f_g(f_l(X)) \rightarrow Y$. The algorithm f_l is a local dimension reduction to capture local data characteristics and f_g is a holistic dimension reduction to gain global data characteristics.

We select f_l and f_g as uniform manifold approximation and projection (UMAP) and principal component analysis (PCA) respectively in the dimension reduction stacking though other options are available theoretically [21]. UMAP is a manifold learning technique producing a low-dimensional embedding by minimizing the fuzzy set cross-entropy between two distributions modeling the similarities of data points in input and embedding spaces [21]. Unlike PCA, it captures local data characteristics well especially when the neighborhood size is set in a relatively small range (e.g., 10). Different from t-distributed stochastic neighbor embedding (t-SNE), it is less random and more explainable by using a 'transparent' neighborhood size rather than a vague perplexity parameter to define a neighborhood and estimate the Gaussian distribution variance for input data [12].

The proposed dimension reduction stacking: UMAP + PCA calculates the UMAP embedding for input data to capture local data characteristics before applying PCA to the embedding to retrieve global data characteristics. It maps the daily excess observations in $\mathbb{R}^{n \times p}$ to the UMAP embedding space $\mathbb{R}^{n \times k}$ before obtaining the PCA embedding in $\mathbb{R}^{n \times l} : p \gg k \geq l$, i.e., $f_{pca}(f_{umap}(X \in \mathbb{R}^{n \times p})) \rightarrow Y \in \mathbb{R}^{n \times l}$. The complexity of the UMAP + PCA stacking is $O(nk^2 + k^3 + n\log n)$. In the implementation, we choose the dimensionality of the UMAP space $k = 10$ and project the UMAP embedding to the subspace spanned by l principal components (PCs) such that they have the explained variance ratio $\eta \geq 90\%$. The input vectorized data is normalized through the minmax normalization for its good performance. Algorithm 1 describes the proposed algorithm.

Algorithm 1: Dimension reduction stacking algorithm (X, nb_{size}, k, η)

Input:

    The input data $X \in \mathbb{R}^{n \times p}$

    The neighborhood size in UMAP: nb_{size}

    The dimensionality of the UMAP embedding: k

    The explained variance ratio: η

Output:

    The dimension reduction stacking embedding: $X_{embedding}$

1.   // UMAP embedding
2. $X_{umap} \leftarrow umap(X, nb_{size}, k)$
3.   //PCA for the UMAP embedding
4. $newData, pcVariance \leftarrow pca(X_{umap})$
5.   //Retrieve the final embedding
6.   **if** $\frac{\sum_{i=1}^{l} pcVariance_i}{\sum_{i=1}^{k} pcVariance_i} \geq \eta$
7. $X_{embedding} \leftarrow newData[:, 1:l]$
8.   **Return** $X_{embedding}$

Different orders of f_{pca} and f_{umap} in stacking can bring totally different dimension reduction results. For example, the proposed UMAP + PCA stacking captures local data characteristics before seeking global ones from the UMAP embedding. The order guarantees that UMAP + PCA can extract local and global data characteristics well in dimension reduction. If PCA goes before UMAP, it will be hard to retrieve local data characteristics because of the dominance of global data characteristics in PCA. The embedding of the PCA + UAMP stacking may lead to overfitting because of the overrepresented global data characteristics in the embedding. Theoretically, f_{pca} or f_{umap} can be substituted by any dimension reduction methods (e.g., $f_{umap} \leftarrow f_{tsne}$), but they may not guarantee to achieve the same stacking results and some inappropriate stacking (e.g., PCA + tSNE) can fail the following forecasting by generating the imbalanced point.

4 Results

The stock excess return forecasting with the 8-K reports is challenging because of the nature of imbalanced learning and the high nonlinearity of data. The mixture of the three types of observations in Fig. 1 reveals it via a visualization approach. Therefore, it can be difficult to achieve high-performance forecasting on such nonlinear data under imbalanced learning and an efficient and robust ML model is needed to tackle the challenge.

We propose a multi-class SVM model to forecast the stock excess returns for its learning efficiency, reproducibility, and interpretability. The one-versus-one ('*ovo*') scheme is employed to extend relevant binary SVM forecasting to corresponding multiclass forecasting to mitigate the impacts of imbalanced data. Unlike deep learning models, SVM owns good interpretability for its simple learning structure and transparent parameter setting. Compared to other ML models such as random forests (RF), it is more reproducible to repeat decent performance because of its rigorous quadratic programming solving. We briefly describe the binary SVM model as follows.

A binary SVM model constructs an optimal hyperplane $y = w^T x + b$ to separate two groups of data points of the training data $X = \{x_i, y_i\}_{i=1}^n, x_i \in R^p, y_i \in \{-1, +1\}$ to by solving a quadratic programming problem:

$$min_w \frac{1}{2} w^T w + C \sum_{i=1}^n \xi_i, w \in R^d, \xi_i \in R, b \in R$$

$$s.t. y_i \left(w^T \varphi(x_i) + b \right) \geq 1 - \xi_i, \xi_i \geq 0, i = 1, 2 \cdots n, \qquad (2)$$

where $w \in R^p$ is the weight vector, $C \in R^+$ is the regularization parameter, and $\varphi(\cdot)$ is an implicit feature function mapping input data to the high-dimensional feature space for evaluation using kernel tricks. The Gaussian kernel $K(x_i x_j) = e^{-\Upsilon \|x_i - x_j\|}, \Upsilon > 0$ is employed in SVM to model nonlinear relationships. The kernel parameter Υ is tuned under the grid search technique for the sake of effective forecasting.

Peer Methods. We employ ML models from shallow learning, midlevel learning, and deep learning as the peer methods of SVM forecasting. Shallow learning refers to those ML models with neither a rigorous mathematical model nor a formal learning topology. For example, k-NN and decision trees (DT) are typical shallow learning methods. Deep learning refers to those with both complex learning topologies and complicated mathematical models. The long short-term memory (LSTM), convolution neural networks (CNN), and transformer models all are typical deep learning methods. Midlevel learning refers to those with complicated mathematical models but less complex learning topologies. SVM, random forests (RF), and other ensemble learning models are typical midlevel learning methods.

We find that SVM generally outperforms its peers in forecasting under the TF-IDF vectorization. SVM achieves the best performance compared to k-NN, RF, and three deep learning models: CNN, LSTM, and Transformer for its highest d-index (1.206). LSTM has the second-best performance with a d-index: 1.187 with an accuracy: 0.48, sensitivity: 0.380, and specificity: 0.697. The k-NN and RF report equivalent performance with d-index: 1.178, but Transformer misclassifies all minority observations as the majority type with a d-index of 1.151 by generating an imbalanced point. It echoes the result of Theorem 1 because of $d = \log_2\left(\frac{3(1+\gamma)}{2}\right)_{\gamma=0.4801} = 1.1506$.

In contrast to the traditional belief that dimension-reduction is 'a must' for vectorized data in NLP, we find that the original data will show advantages over the reduced data from dimension reduction in forecasting under the TF-IDF vectorization. Table 1 compares the SVM forecasting on the original data and different reduced data in terms of the d-index and classic measures such as accuracy, sensitivity, specificity, precision, and NPR (negative prediction ratio). It shows that SVM can forecast the stock excess returns better using the original data than the reduced data. For example, SVM forecasting achieves a 1.206 d-index on the original data but only a d-index of 1.162 on the reduced data from PCA (SVM_{pca}). This suggests that dimension reduction may cause some information missing so that the signal-to-noise (SNR) ratio of the reduced data decreases. Moreover, SVM forecasting with PCA + UMAP stacking ($SVM_{pca+umap}$) has almost the same performance as SVM_{pca}. It implies the holistic dimension reduction algorithm PCA may dominate the feature extraction in the PCA + UMAP stacking.

Table 1. The SVM excess return forecast under TF-IDF

Measures	SVM	SVM_{pca}	$SVM_{pca+tsne}$	$SVM_{pca+umap}$	$SVM_{umap+pca}$
D-index	1.206	1.162	1.150	1.161	1.186
Accuracy	0.496	0.484	0.479	0.483	0.482
Sensitivity	0.384	0.342	0.333	0.343	0.376
Specificity	0.699	0.673	0.667	0.672	0.694
Precision	0.425	0.406	Nan	0.440	0.398
NPR	0.740	0.756	Nan	0.741	0.724

However, the proposed UMAP + PCA outperforms PCA, PCA + UMAP, and PCA + tSNE in SVM forecasting because of the 1.186 d-index value of $SVM_{umap+pca}$. The

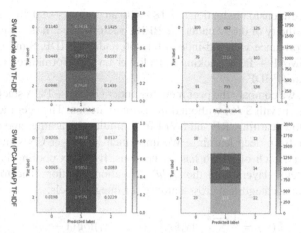

Fig. 2. The normalized and general confusion matrices of the SVM and SVM$_{pca+umap}$ forecasts under the TF-IDF vectorization. The SVM$_{pca+umap}$ forecast is affected more by the majority type ('hold') because 96.58% of 'buy' and 95.74% of 'sell' observations are misclassified as the majority type respectively. However, the corresponding ratios of the SVM forecasting on the original data are 74.34% and 76.20%.

SVM$_{pca}$, SVM$_{pca+umap}$ and SVM$_{pca+tsne}$ forecasting all generate the imbalanced points or AIPs because their d-index values are close to 1.1508, the d-index value at the imbalanced point. It seems that the inappropriate stacking: PCA + tSNE forces SVM to lose its learning capability by generating the imbalanced point in forecasting for its d-index of 1.150. It is possible that the global data characteristics are over-retrieved in the PCA + tSNE stacking which may cause the Gaussian kernel in SVM to lose its discriminability. On the other hand, the better performance of SVM$_{umap+pca}$ may lie in the built-in advantage of UMAP + PCA in retrieving both global and local data characteristics to avoid the dominance of global data characteristics in dimension reduction. The UMAP + PCA stacking can bring almost equivalent forecasting besides greatly decreasing the complexity of foresting, by producing well-defined reduced data for high-dimensional vectorized data.

Figure 2 compares the normalized and general confusion matrices of the SVM and SVM$_{pca+umap}$ forecasts under the TF-IDF vectorization. It shows that SVM without dimension reduction demonstrates a better excess return forecast than SVM$_{pca+umap}$. The SVM$_{pca+umap}$ forecast is hijacked more by the majority type ('hold') because 96.58% of 'buy' and 95.74% of 'sell' observations are misclassified as the majority type respectively. However, the corresponding ratios of the SVM forecast without dimension reduction are 74.34% and 76.20%.

4.1 Dimension Reduction Stacking Under the BERT Vectorization

Unlike the general assumption that BERT is superior to TF-IDF in text vectorization, we find the SVM forecast shows advantages under the TF-IDF vectorization in comparison to the BERT vectorization for the 8-K data [14]. For example, the best d-index achieved by SVM without dimension reduction is 1.189 with an accuracy of 0.480, a sensitivity of 0.384, and a specificity of 0.699 under the BERT vectorization. It is slightly lower than the forecast performance with a d-index of 1.206 under TF-IDF. Similarly, The SVM forecast under UMAP + PCA stacking has a lower-level performance under the BERT vectorization than TF-IDF.

Furthermore, we compare the peer methods under UMAP + PCA stacking under the BERT vectorization with SVM without dimension reduction. The peer methods include k-NN from shallow learning, RF and SVM from midlevel learning, and CNN, LSTM, and Transformer from deep learning. We implement a 7-layer CNN model with two convolution layers, each of which has 64 neurons, two max-pooling layers, a flatten layer, and two dense layers with the *'relu'* activation function. The loss function is chosen as the sparse categorical cross-entropy for integer labels,

$$L(w) = \frac{1}{n}\sum\nolimits_{i=1}^{n}[y_i log\hat{y}_i + (1 - y_i)log(1 - \hat{y}_i)] \tag{3}$$

where w represents the weights of the network, y_i and $\hat{y}_i$, represent the true label and predicted label respectively.

We also implement a 6-layer LSTM model with 3 LSTM layers and 2 dense layers. The LSTM layers have [128, 64, 64] neurons with 'Tanh' activation functions and the dense layers have [128, 3] neurons with the *'relu'* and *'softmax'* activation functions. The loss function is selected as the sparse categorical cross-entropy. The drop rates increase gradually from 0.1 to 0.3 with the depth of the layers. In addition, we implement a transformer model with 4 dense layers and an embedding layer with the sparse categorical cross-entropy loss function. The drop rates are chosen as 0.2 generally.

Figure 3 compares the SVM forecast without dimension reduction with the peer methods with the UAMP + PCA stacking under the BERT vectorization in terms of classic classification metrics (left plot) and d-index values (right plot). The shallow learning model k-NN and the midlevel learning model RF under UMAP + PCA achieve the best and second-best performance, but the deep learning models only achieve mediocre performance. In particular, the Transformer model generates an imbalanced point by misclassifying all minority observations in the test data as the majority type. This suggests that the attention mechanism of the transformer model may not work well under imbalanced data because it may amplify the impact of the majority type in learning [22]. Therefore, the sophisticated deep learning models may suffer from their built-in weakness in encountering overfitting for their complicated topologies compared to the shallow and midlevel learning models.

However, the best peer (e.g., RF) under the UMAP + PCA stacking may still fall behind the SVM model without dimension reduction. It echoes the previous result that dimension reduction may not contribute to enhancing forecasting even though vectorized data with high dimensionality.

4.2 Stock Excess Return Forecast Under Imbalance Resampling

Since the stock excess return prediction is essentially an imbalanced learning problem, we employ resampling techniques: random oversampling (ROS) and Tomek links (TL)

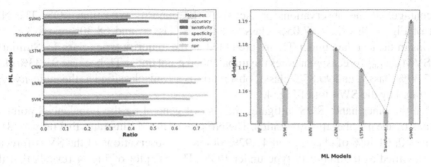

Fig. 3. The comparisons of peer methods under the UMAP + PCA stacking with the SVM forecast without dimension reduction (SVM0) under the BERT vectorization. The SVM forecast outperforms the peers with the stacking. The deep learning models only have mediocre or even much poor performance compared to k-NN and RF under the stacking for its complicated topologies.

to handle data imbalance. ROS is an oversampling method to increase the number of observations of the minority type(s) and TL is an under-sampling method to decrease the number of majority-type observations to achieve data balance. We choose TF-IDF vectorized data for its advantage in forecasting.

We find the resampling techniques generally cannot enhance the SVM forecast. Instead, ROS mitigates the risk of overfitting by increasing misclassification ratios. TL tends to increase the risk of overfitting and generate the imbalanced point. Table 2 compares the performance of the SVM and $SVM_{pca+tsne}$ under TL and ROS. SVM has obvious advantages over $SVM_{pca+tsne}$, which generates the imbalanced point, under the two resampling methods in terms of the d-index values. The SVM forecast under TL stands at the same level as the original SVM forecast for their close d-index values: 1.210 and 1.206, but the SVM forecast under ROS only achieves a d-index of 1.175. It suggests that the ROS resampling would bring more noise into forecasting through the oversampling procedure and decrease the effectiveness of prediction.

Table 2. The SVM excess return forecast with resampling under TF-IDF

Measures	SVM (TL)	SVM (ROS)	$SVM_{pca+tsne}$ (TL)	$SVM_{pca+tsne}$ (ROS)
D-index	1.210	1.175	1.160	1.149
Accuracy	0.492	0.454	0.478	0.438
Sensitivity	0.395	0.400	0.347	0.384
Specificity	0.705	0.708	0.676	0.700
Precision	0.426	0.402	Nan	0.382
NPR	0.735	0.712	0.721	0.703

Figure 4 compares the confusion matrices of the SVM and $SVM_{pca+tsne}$ forecasts under TL and ROS. It shows that the SVM forecast under TL tends to increase the percentages of the observations predicted as the majority type than ROS. The NW plot of Fig. 4 illustrates 67.05% class 0, 84.86% class 2, and 68.30% class 3 observations in the test data under TL are predicted as the majority type (class 1). Similarly, the $SVM_{pca+tsne}$ forecast generates the imbalanced point under TL because 91.33% class 0, 95.39% class 1, and 90.85% class 2 observations are misclassified as the majority type illustrated by the SW plot of Fig. 4.

On the other hand, ROS mitigates the risk of generating the imbalanced point by increasing the misclassification ratios. The NE plot of Fig. 4 illustrates that only 44.81% of class 0, 64.46% of class 1, and 47.92% of class 2 observations in the SVM forecast are classified as the majority type under ROS. The SE plot of Fig. 4 reveals that the overfitting of the $SVM_{pca+tsne}$ forecast is decreased under ROS.

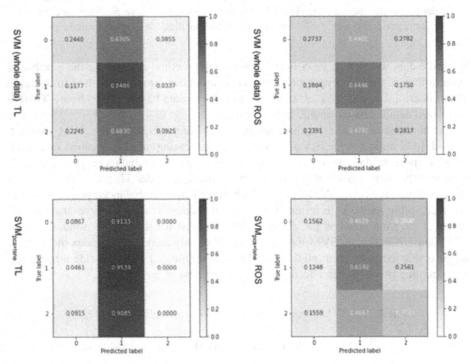

Fig. 4. The normalized confusion matrix comparisons of the SVM and $SVM_{pca+tsne}$ forecasts under the random oversampling (ROS) and Tomek links (TL) resampling techniques. TL tends to increase the percentages of the observations in all classes predicted as the majority type. ROS mitigates the overfitting risk of generating the imbalanced point by increasing more misclassifications.

5   Discussion and Conclusion

We model the stock excess forecast using the SEC 8-K filings as an imbalanced learning problem and propose an SVM model with tuned Gaussian kernels to achieve applicable forecasting in business practice. Besides good reproducibility, the proposed model demonstrates favorable performance compared to the peers. We also find that dimension reduction may not contribute to improving forecast effectiveness as expected compared to using the original data. Instead, they may lower the forecast performance slightly though they decrease the learning complexities in the forecast. In addition, Inappropriate dimension reduction (e.g., PCA + tSNE) can increase the risk of overfitting in the forecast or even cause the ML model to lose learning capabilities. It also shows that resampling techniques cannot enhance the forecasting effectiveness. The TL resampling tends to increase the risk of overfitting in forecasting, but the ROS resampling can mitigate the risk of overfitting by increasing misclassifications.

We also propose a novel dimension reduction stacking method: UMAP + PCA to retrieve both global and local data characteristics for vectorized data. It outperforms other dimension reduction methods in SVM forecasting and can even substitute the SVM forecast to avoid the huge forecasting complexity from high-dimensional vectorized data. Besides, we find that the TF-IDF vectorization would demonstrate some advantage over the BERT vectorization in the stock excess return forecast.

Some questions remain to be answered on this topic. For example, how to enhance forecasting by minimizing the impact from the majority type? Since the classic resampling methods fail on this imbalanced learning problem, it is possible to employ GAN models to generate minority samples to balance data to seek an alternative solution [23]. Furthermore, we are using sparse coding techniques to conduct post-processing for input data or reduced data from the dimension reduction stacking to improve forecasting besides investigating the integration of classic Fourier analysis with the novel stacking technique to extract more meaningful features to attain high-performance forecasting [24]. It is also possible to integrate the stacking technique with interpretable deep learning models to address the high nonlinearity of the 8-K data [25]. To the best of our knowledge, the algorithms and techniques developed in this study will not only help stakeholders to optimize their investment decisions by exploiting the SEC 8-K filings but also shed light on AI innovations in digital business.

Acknowledgements. This work is partially supported by McCollum endowed chair startup fund.

References

1. Loughran, T., McDonald, B.: Textual analysis in accounting and finance: a survey. J. Account. Res. **54**(4), 1187–1230 (2016)
2. Xie, B., et al.: Semantic frames to predict stock price movement. In: Proceedings of the 51st Annual Meeting of the Association for Computational Linguistics, pp. 873–883 (2013)
3. Ke, Z., Kelly, B., Xiu, D.: Predicting returns with text data (No. w26186). National Bureau of Economic Research (2019)
4. Loughran, T., McDonald, B.: When is a liability not a liability? Textual analysis, dictionaries, and 10-Ks. J. Financ. **66**(1), 35–65 (2011)
5. Zhai, S., Zhang, Z.: Forecasting firm material events from 8-K reports. In: Proceedings of the Second Workshop on Economics and Natural Language Processing, pp. 22–30 (2019)
6. Kogan, S., et al.: Predicting risk from financial reports with regression. In: Proceedings of human language technologies: the 2009 annual conference of the North American Chapter of the Association for Computational Linguistics, pp. 272–280 (2009)

7. Lee, H., et al.: On the importance of text analysis for stock price prediction. LREC **2014**, 1170–1175 (2014)
8. Zhao, X.: Does information intensity matter for stock returns? Evidence from Form 8-K filings. Manage. Sci. **63**(5), 1382–1404 (2017)
9. Engelberg, J.: Costly information processing: evidence from earnings announcements. AFA 2009 San Francisco meetings paper (2008)
10. Li, F.: The information content of forward-looking statements in corporate filings, a naïve Bayesian machine learning approach. J. Account. Res. **48**(5), 1049–1102 (2010)
11. Aydogdu, M., et al.: Using long short-term memory neural networks to analyze SEC 13D filings: a recipe for human and machine interaction. Intelligent Systems in Accounting, Finance and Management (2020). https://doi.org/10.1002/isaf.1464
12. Han, H., et al.: Enhance explainability of manifold learning. Neurocomputing **500**, 877–895 (2022)
13. Lee, S.: Document vectorization method using network information of words. PLoS ONE **14**(7), e0219389 (2019). https://doi.org/10.1371/journal.pone.0219389
14. Devlin, J., et al.: BERT: Pre-training of Deep Bidirectional Transformers for Language Understanding. arXiv:1810.04805 (2019)
15. Lansing, K., LeRoy, S., Ma, J.: Examining the Sources of Excess Return Predictability: Stochastic Volatility or Market Inefficiency? Federal Reserve Bank of San Francisco Working Paper (2018)
16. Cristianini, N., Shawe-Taylor, J.: An Introduction to Support Vector Machines and Other Kernel-based Learning Methods. Cambridge University Press (2000)
17. Han, H.: Hierarchical learning for option implied volatility pricing. In: Hawaii International Conference on System Sciences (2021)
18. NLP-for-8K-documents. https://github.com/hatemr/NLP-for-8K-documents
19. Han H., et al.: Interpretable Machine Learning Assessment (2022). Available at SSRN: https://ssrn.com/abstract=4146556
20. Van der Maaten, L., Hinton, G.: Visualizing data using t-SNE. J. Mach. Learn. Res. **9**(11) (2008)
21. McInnes, L., Healy, J., Melville, J.: UMAP: Uniform Manifold Approximation and Projection for Dimension Reduction. arXiv:1802.03426. (2020)
22. Vaswani, A., et al.: Attention is All you Need, NIPS (2017)
23. Goodfellow, I.J., et al.: Generative adversarial nets. Neural Information Processing Systems, NIPS, pp 2672–2680 (2014)
24. Han, H., Jiang, X.: Overcome support vector machine diagnosis overfitting. Cancer Inform. **13**(1), 145–158 (2014)
25. Gas, R., et al.: Explainable Deep Learning: A Field Guide for the Uninitiated (2021). https://doi.org/10.48550/arXiv.2004.14545

A Fast Initial Response Approach to Sequential Financial Surveillance

Michael Pokojovy$^{(\boxtimes)}$ and Andrews T. Anum

Department of Mathematical Sciences, The University of Texas at El Paso, El Paso, TX 79968, USA
mpokojovy@utep.edu, atanum@miners.utep.edu

Abstract. Consider the problem of financial surveillance of a heavy-tailed time series modeled as a geometric random walk with log-Student's t increments assuming a constant volatility. Our proposed sequential testing method is based on applying the recently developed taut string (TS) univariate process monitoring scheme to the gaussianized log-differenced process data. With the signal process given by a properly scaled total variation norm of the nonparametric taut string estimator applied to the gaussianized log-differences, the change point detection procedure is constructed to have a desired in-control (IC) average run length (ARL) assuming no change in the process drift. If a change in the process drift is imminent, the proposed approach offers an effective fast initial response (FIR) instrument for rapid yet reliable change point detection. This framework may be particularly advantageous for protection against imminent upsets in financial time series in a turbulent socioeconomic and/or political environment. We illustrate how the proposed approach can be applied to sequential surveillance of real-world financial data originating from Meta Platforms, Inc. (FB) stock prices and compare the performance of the TS chart to that of the more prominent CUSUM and CUSUM FIR charts at flagging the COVID-19 related crash of February 2020.

Keywords: Geometric random walk · Heavy tails · Taut string (TS) chart · CUSUM FIR chart · FB stock

1 Introduction

Financial meltdowns have raided financial markets numerous times over the past decades. Famous examples of such crashes include, but are not limited to, the crash following the burst of the United States housing market bubble in the

This work was funded partially by the National Institute on Minority Health and Health Disparities (NIMHD) grant (U54MD007592). Comments and improvement suggestions from SDSC 2022 General Chair Professor H. Han and anonymous referees are greatly appreciated.

H. Han and E. Baker (Eds.): SDSC 2022, CCIS 1725, pp. 19–33, 2022.
https://doi.org/10.1007/978-3-031-23387-6_2

1990s, the 2000 dot-com bubble, the Dow Jones Industrial Average crash associated with the COVID-19 pandemic in 2020 and, very recently, the after-hours crash of Facebook Inc. (FB) stock in February 2022 or Netflix (NFLX) stock in April 2022. For all of these abrupt drops in stock prices or price-weighted indices, the underlying problem is similar whether the precipitous price change was expected or unanticipated, invariably leading to severe financial and economic consequences. In situations where crashes or upsets are expected to happen in a near future, the question of when exactly the crash is going to happen still remains a challenging and non-trivial one.

Financial surveillance [9] and predicting crashes [5,15,16] are two closely related yet independent topics. While the latter one aims at predicting the risk of future crashes over a given time horizon, financial surveillance is meant to flag a crash as soon as it starts to manifest itself. Thus, we would like to emphasize that the methodology presented in this paper is not meant to serve as a replacement for forecasting future crashes, but rather offer an extra layer of protection against situations when crash prediction fails.

To put this into a formal framework, consider the problem of financial surveillance of a heavy-tailed time series modeled as a geometric random walk with log-Student's t increments under the assumption of a constant volatility. If the process drift changes, future paths of such geometric random walk may have severe structural breaks (unusual behavior) which can have major financial and economic ramifications for individual investors, particular markets or the economy as a whole. It is not easy, if at all possible, to exactly tell when this break(s) will occur. It is, therefore, of much importance for financial advisors, stock traders, investors, etc., to have a statistical procedure that is consistently able to early detect unusual behaviors of financial time series.

This problem is encountered not only in financial surveillance but also transcends to areas such as industrial process monitoring, public health and computer network monitoring [9,10,20], etc. Numerous attempts to solve the statistical problem of early change-point detection have been made over the past decades. The earliest developments date back to the popular Shewhart X chart and the Cumulative Sum (CUSUM) stopping rule to sequentially monitor an independent Gaussian process data. The Shewhart X chart as well as the CUSUM have been widely used in detecting anomalous patterns because of their optimality properties (in a suitable sense) and, even more so, due to the simplicity of the methods. The intrinsic problem associated with these methods comes to bare when the focus is rather on monitoring processes with non-constant mean shifts. Also, unlike Shewhart X chart, the CUSUM technique expects sustained shift in the mean and also requires knowledge about the magnitude of the shift, i.e., is not reference-free. Additionally, both methods are not optimal in fast initial response (FIR) situations. Indeed, the CUSUM initial signal is set zero at start-up so that the chart requires some time to gain its full efficiency. The Shewart X chart, being memoryless, cannot be tuned to rapidly respond to initial-state upsets. Consequently, these classic charts may perform suboptimally when a change-point is imminent. The FIR feature due to its advantage of reducing the average detection time (in case the upset happens in the initial state) has been currently adopted and applied to several statistical process control methodologies such as the FIR CUSUM in [14]. Variants of the CUSUM chart have been used in [3,6,8] to detect upsets in financial time series but under different conditions.

We propose a sequential testing method based on applying the recently developed taut string (TS) univariate process monitoring scheme [21] to gaussianized log-differenced process data. With the signal process given by the total variation of the nonparametric taut string estimator applied to the latter gaussianized log-differences, the change point detection procedure is constructed to have a desired in-control (IC) average run length (ARL) assuming no change in process drift. Assuming a change in the process drift is imminent, the proposed approach offers an effective fast initial response (FIR) instrument for rapid yet reliable change point detection. This was demonstrated in [21] for independent process data based on extensive simulations for a wide variety of steady state sizes and sustained upset magnitudes. Extending this approach to a class of heavy-tailed time series data, the resulting framework may be particularly advantageous for protection against imminent upsets in financial time series in turbulent socioeconomic and/or political environments. Adopting the original TS chart from [21], our proposed method does not require any prior knowledge about the type or magnitude of shift in the mean so that the procedure can be described as a reference-free technique. We illustrate how the proposed approach can be applied to sequential surveillance of real-world financial data originating from Meta Platforms, Inc. (FB) stock prices and compare the performance of the TS chart to that of the more prominent CUSUM and CUSUM FIR charts at detecting the recent crash of February 2020.

2 Geometric Random Walk with Log-Student's t Increments

A widely adopted model to describe the evolution of a risky asset price $(S(t))_{t\geq 0}$ is given by stochastic differential equation (SDE)

$$\mathrm{d}S(t) = rS(t)\mathrm{d}t + \sigma S(t)\mathrm{d}W(t), \quad S(0) = S_0, \tag{1}$$

where S_0 is a given initial value and $(W(t))_{t\geq 0}$ is a standard Wiener process. The increments thereof, known as the "white noise," supply innovations for this time series model. The scalar parameters, $r \in \mathbb{R}$ and $\sigma > 0$, are referred as drift and volatility. Being a basic heteroscedastic continuous-time model, Eq. (1) is widely applied in quantitative finance [7]. Recently, it was also applied to model and predict biomedical time series [2]. On the strength of Itô's calculus, the solution process $(S(t))_{t\geq 0}$ is given by

$$S(t) = S_0 \exp\left\{\left(r - \frac{1}{2}\sigma^2\right)t + \sigma W(t)\right\}. \tag{2}$$

Assume the stochastic process $(S(t))_{t \geq 0}$ is observed over an equispaced time grid $\{t_0, t_1, \ldots, t_n\}$ for some constant step size $\Delta t \equiv t_k - t_{k-1} > 0$. Computing the log-differences

$$x(t_k) := \ln(S(t_k)) - \ln(S(t_{k-1})) = \left(r - \frac{1}{2}\sigma^2\right)\Delta t + \sigma\left(W(t_k) - W(t_{k-1})\right),$$

we obtain

$$x(t_k) \overset{\text{i.i.d.}}{\sim} \mathcal{N}\left((\Delta t)\left(r - \frac{1}{2}\sigma^2\right), (\Delta t)\sigma^2\right). \tag{3}$$

We refer to the process $(S(t_k))_{t_k \geq 0}$ as in-control (IC) if the process drift $r(t) \equiv r_{\text{IC}}$ remains constant over time. Assuming the volatility σ is constant, on the strength of Eq. (3), this is equivalent to the process mean of the $(x(t_k))_{t_k \geq 0}$ being constant.

Empirically [19,22], one can commonly observe the increments exhibit a heavier tail than the Gaussian one. Thus, the Gaussianity in Eq. (3) is often replaced with the more adequate assumption of i.i.d. Student's t log-increments

$$x(t_k) \overset{\text{i.i.d.}}{\sim} t_\nu\left((\Delta t)\left(r - \frac{1}{2}\sigma^2\right), (\Delta t)\sigma^2\right) \tag{4}$$

for some $\nu \in (0, \infty]$ with the limiting case $\nu = \infty$ corresponding to the original case of log-Gaussian increments.

Letting $\hat{\mu}_x$ and $\hat{\sigma}_x$ denote location and scale estimates obtained from $x(t_k)$'s, e.g., $\bar{x}$ and $\sqrt{\frac{\nu-2}{\nu}s^2}$ or their robust MCD counterparts, Eq. (4) implies

$$\hat{\mu}_x = (\Delta t)\left(\hat{r} - \frac{1}{2}\hat{\sigma}^2\right) \quad \text{and} \quad \hat{\sigma}_x^2 = (\Delta t)\hat{\sigma}^2$$

or, solving for $\hat{r}$ and $\hat{\sigma}$, the drift and volatility can be estimated via

$$\hat{r} = \frac{\hat{\mu}_x}{\Delta t} + \frac{1}{2}\frac{\hat{\sigma}_x^2}{\Delta t} \quad \text{and} \quad \hat{\sigma} = \frac{\hat{\sigma}_x}{\sqrt{\Delta t}}.$$

Since most statistical process monitoring (SPM) methodologies assume Gaussian data and most robust estimation procedures are calibrated for the Gaussian distribution, it is convenient to gaussianize the log-differences $x(t_k)$ before estimating IC standards and applying an SPM chart. If the degrees of freedom ν are known, we can transform

$$y(t_k) = F_{t_\infty}^{-1}\left\{F_{t_\nu}\left(\frac{x(t_k) - \left(r - \frac{1}{2}\sigma^2\right)}{\sqrt{\Delta t}\sigma}\right)\right\}$$

to render the log-differences standard Gaussian where $F_{t_\nu}(\cdot)$ is the cdf of a standard t_ν-random variable. (Note $F_{t_\infty}(\cdot)$ is the standard Gaussian cdf.) Since the IC drift and volatility may be unknown, a possible empirical alternative is

$$y(t_k) = F_{t_\infty}^{-1} \left\{ F_{t_{\hat{\nu}}} \left(\frac{t_{\hat{\nu}, 0.75} - t_{\hat{\nu}, 0.25}}{\widehat{\mathrm{IQR}}(\{x(t_k) \mid k = 1, 2, \ldots, n\})} \right. \right.$$

$$\left. \left. \times \left(x(t_k) - \widehat{\mathrm{median}}(\{x(t_k) \mid k = 1, 2, \ldots, n\}) \right) \right) \right\}.$$

Despite the fact the $y(t_k)$'s are expected to be approximately i.i.d. standard Gaussian, instead of using 0 and 1 as IC standards for process mean and standard deviation, we recommend to employ robust statistical estimators with a high breakdown point and good efficiency properties to estimate the IC standards of the $y(t_k)$-process. This offers an effective way to safeguard against outliers and possible model assumption violations. Namely, we employ the MCD and the MDPD estimators discussed in Sect. 3.

3 Robust Model Calibration

Many classical statistical estimators suffer from severe bias effects in the presence of outliers in the dataset. Consequently, robust alternatives, i.e., estimators that are not too sensitive to outliers, have been put forth. Examples of such robust location and scale estimators are the minimum density power divergence (MPDP) and minimum covariance determinant (MCD). A thorough comparison of statistical properties (such as empirical breakdown point and efficiency) of these estimators was performed in [2]. We use a `Matlab`® implementation of MCD and MDPD estimator available at https://github.com/AndrewsJunior/mdpd1D.

3.1 Minimum Covariance Determinant (MCD)

The raw minimum covariance determinant (MCD) estimator [11] is a well-known location and scatter estimator. We present the univariate version thereof. Given a univariate sample dataset $\{x_i, x_2, \ldots, x_n\}$ and a desired "best set" size $\lfloor \frac{n}{2} \rfloor + 1 \leq h \leq n$, the MCD location and scale estimator pair $(\hat{\mu}_{\mathrm{MCD}}, \hat{\sigma}^2_{\mathrm{MCD}})$ is computed as follows.

Order the observations x_i in an increasing manner, i.e., $x_{(1)} \leq x_{(2)} \leq \ldots \leq x_{(n)}$, and consider the following $(n - h + 1)$ contiguous subsamples:

$$\{x_{(1)}, \ldots, x_{(h)}\}, \{x_{(2)}, \ldots, x_{(h+1)}\}, \{x_{(3)}, \ldots, x_{(h+2)}\}, \ldots, \{x_{(n-h+1)}, \ldots, x_{(n)}\}.$$

For each of these subsamples $\{x_{(i)}, \ldots, x_{(h+i-1)} | i = 1, \ldots, n - h + 1\}$ of size h, the sample mean and the corresponding sum of squares are computed as follows:

$$\mu^{(1)} = \frac{1}{h} \sum_{i=1}^{h} x_{(i)}, \quad \mu^{(2)} = \frac{1}{h} \sum_{i=2}^{h+1} x_{(i)}, \quad \ldots, \quad \mu^{(n-h+1)} = \frac{1}{h} \sum_{i=n-h+1}^{n} x_{(i)} \quad \text{and}$$

$$\sigma^{(1)} = \sum_{i=1}^{h} \left(x_{(i)} - \mu^{(1)} \right)^2, \quad \ldots, \quad \sigma^{(n-h+1)} = \sum_{i=n-h+1}^{n} \left(x_{(i)} - \mu^{(n-h+1)} \right)^2.$$

The MCD scatter estimate is then given as

$$\hat{\sigma}_{\mathrm{MCD}}^2 = k_{\mathrm{MCD}} \cdot c_{\mathrm{MCD}} \cdot \min_{1 \le j \le n-h+1} \sigma^{(j)}$$

where the asymptotic bias correction factor $k_{\mathrm{MCD}} = (h/n)/\mathbb{P}(\chi_3^2 \le \chi_{1,h/n}^2)$ and c_{MCD} is a small sample bias correction factor implemented in the (internal) `corfactorRAW()` function of the `robustbase` R-package. Continuing, the MCD location estimate, $\hat{\mu}_{\mathrm{MCD}}$, is the mean $\mu^{(j)}$ associated with the optimal $j^* = \arg\min_{1 \le j \le n-h+1} \sigma^{(j)}$. Obtaining the univariate MCD location and scale estimates in this fashion is computationally expensive since one has to run through each subsample two times. A more computationally efficient way of obtaining these estimates is by using recursive formula presented in [23]. The location and scale estimates are computed recursively for $j > 1$ by

$$\mu^{(j)} = \frac{h\mu^{(j-1)} - x_{(j-1)} + x_{(j+h-1)}}{h} \text{ and}$$

$$\sigma^{(j)} = \sigma^{(j-1)} - (x_{(j-1)})^2 + (x_{(j+h-1)})^2 - h(\mu^{(j)})^2 + h(\mu^{(j-1)})^2,$$

respectively. Both $\mu^{(j)}$ and $\sigma^{(j)}$ use estimates from the previous subsample. Again, once all $\sigma^{(j)}$'s are available, the MCD location and scatter estimates are calculated as described above. See [23, Chapter 4, Sect. 3] for the algorithm. The MCD estimator has attractive statistical properties such as high breakdown point, affine equivariance, high efficiency and bounded influence function.

3.2   Minimum Density Power Divergence (MDPD)

The robust minimum density power divergence estimator for general parametric models was introduced in [4]. Focusing on univariate distributions, the divergence functional $d_\alpha(f, g)$ between two univariate probability density functions $g(x)$ and $f(x)$ is defined as

$$d_\alpha(f,g) := \int_{\mathbb{R}} \left(f^{1+\alpha}(x) - \left(1 + \frac{1}{\alpha}\right) g(x) f^\alpha(x) + \frac{1}{\alpha} g^{1+\alpha}(x) \right) dx \qquad (5)$$

for positive α. Assuming $x_1, \ldots, x_n$ are independently sampled from $g(x)$, the law of large numbers suggests

$$\int_{\mathbb{R}} f^\alpha(x) g(x) dx \approx \frac{1}{n} \sum_{i=1}^{n} f^\alpha(x_i).$$

The last term in Eq. (5) being independent of $f(x)$, the minimization of $d_\alpha(f(\cdot|\theta), g(\cdot))$ over $\theta \in \Theta$ is asymptotically equivalent with minimizing the density power divergence (DPD) function

$$H_n(\theta) = \int_{\mathbb{R}} f^{1+\alpha}(x|\theta) dx - \left(1 + \frac{1}{\alpha}\right) n^{-1} \sum_{i=1}^{n} f^\alpha(x_i|\theta). \qquad (6)$$

Thus, the minimum density power divergence (MDPD) estimator is given as

$$\hat{\theta}_n^{\mathrm{DPD}} := \arg\min_{\theta\in\Theta} H_n(\theta). \tag{7}$$

The MDPD estimator was shown [4] to exhibit a number of attractive statistical properties such as asymptotic normality, affine equivariance (when estimating location and scale), high breakdown point as well as high efficiency, rendering it very useful for robust inference. A numerical optimization algorithm for solving the minimization problem (7) was recently proposed in [2].

4 Statistical Process Monitoring

Statistical process monitoring (SPM), also referred to as statistical process control or process surveillance, involves sequential hypothesis testing for a quantity of interest sampled at regular time intervals. In quality engineering, this is of necessity because defective items can be produced during the manufacturing process. If an efficient quality control routine is performed at the end of that production, some, if not all, of these defective items may be detected. One major disadvantage of monitoring the process at the end of production is that, in many cases, these defective items cannot be salvaged or repaired. In contrast, we consider process monitoring techniques that are performed sequentially and have the capacity to warn the analyst if the process appears to be out-of-control (OC) or, in terms of financial surveillance, whether a change in the stock price drift has occurred. We present a concise formal framework for SPM closely following [21].

4.1 Taut String (TS) Fast Initial Response (FIR) Chart

Let $X_1, X_2, \ldots, X_n, \ldots$ be independent random variables observed sequentially at equispaced periods of time. Classical statistical process monitoring methodologies adopt the sequential testing framework

$$H_0 : X_n \overset{\text{i.i.d.}}{\sim} \mathcal{N}(\mu_{\mathrm{IC}}, \sigma_{\mathrm{IC}}^2) \text{ for all } n \in \mathbb{N}, \qquad \text{vs.} \tag{8}$$

$$H_1 : X_n \overset{\text{i.i.d.}}{\sim} \mathcal{N}(\mu_{\mathrm{IC}}, \sigma_{\mathrm{IC}}^2) \text{ for } n \leq \vartheta-1 \text{ and } X_n \overset{\text{i.i.d.}}{\sim} \mathcal{N}(\mu_{\mathrm{OC}}, \sigma_{\mathrm{IC}}^2) \text{ for } n \geq \vartheta \tag{9}$$

for some unknown change point $\vartheta < \infty$. This corresponds to a sustained shift in Gaussian location assuming a constant variance. Without loss of generality, assume $\mu_{\mathrm{IC}} = 0$, $\sigma_{\mathrm{IC}} = 1$ and $\mu_{\mathrm{OC}} = \delta$. In practice, both upset time ϑ and upset magnitude δ are unknown. We focus on two-sided control charts that monitor shifts of size $\pm\delta$ for $\delta > 0$.

The upset time ϑ is assumed deterministic yet unknown. Under the null-hypothesis H_0, the process $(X_n)_{n\in\mathbb{N}}$ is called stable or in-control (IC). The alternative hypothesis H_1 represents a sustained upset beginning at the ϑ-th time period leading to an unstable or out-of-control (OC) process.

Let τ be a non-randomized stopping time. Usually, τ is defined as the first time a signal process $(S_n)_{n \in \mathbb{N}}$ arising from the control chart methodology employed exceeds some given control limit $L > 0$. If the limit L is never exceeded, $\tau := \infty$ is set.

The conventional IC average run length (ARL) and OC ARL (or average detection delay) are given as

$$\text{IC ARL} = \mathbb{E}_\infty[\tau] \text{ and OC ARL}_\vartheta = \mathbb{E}_\infty[(\tau - \vartheta + 1)^+]/\mathbb{P}_\vartheta(\{\tau \geq \vartheta\})$$

where the expectations are taken with respect to probability measures $\mathbb{P}_\infty$ and $\mathbb{P}_\vartheta$ associated with the hypotheses in Eqs. (8) and (9), respectively. The number $(\vartheta - 1)$ represents the steady-state size. In particular, for $\vartheta = 1$, we obtain the initial state. If the alternative hypothesis H_1 holds, it is desirable that a chart detects the upset as quickly as possible. Comparing two stopping rules with equal IC ARLs, the chart with a smaller OC ARL$_\vartheta$ is preferred.

The CUSUM chart (if properly tuned) is known to be optimal in a certain minimax sense described by [12] and [17]. For a given IC ARL level, this optimality is attained assuming $k = \delta/2$ (for a known shift size δ) over long steady states $\vartheta \to \infty$. In practice, high intensity of upsets (short interarrival times between two successive upsets) may prevent existence of long steady states. In this situation, it is recommendable to employ the Fast Initial Response (FIR) approach. An FIR modification of the classical CUSUM chart was proposed in [13] by giving the usual CUSUM signals a "head start" by initializing the upper and lower signal processes as $U_0 = h/2$ and $L_0 = h/2$, where $h > 0$ is a control limit computed to produce a desired IC ARL.

Recently, a novel FIR chart based on nonparametric taut string estimation was developed in [21]. In contrast to CUSUM FIR, this chart does not require prior knowledge of the shift size δ but rather employs nonparametric regression to estimate the latter. Nonparametric estimation makes the TS chart reference free, i.e., capable of detecting non-sustained shifts such as linear drifting or period cycling, etc.

Table 1. Characteristics of TS chart for an IC process (reproduced from [21]).

Taut string (TS) chart	Nominal IC ARL		
	100	370	750
Run length std. dev. ($s_{\text{IC RL}}$)	447.99	1928.89	4078.65
Std. dev. of $\widehat{\text{IC ARL}}$ ($s_{\widehat{\text{IC ARL}}}$)	0.4480	1.9289	4.0786
$\text{IC ARL} + 3s_{\widehat{\text{IC ARL}}}$	101.3440	375.7867	762.2359
Estimated control limit for $\text{IC ARL} + 3s_{\widehat{\text{IC ARL}}}$	1.7706	2.1233	2.3261
Estimated control limit for IC ARL	1.7673	2.1193	2.3214

Let $x_1, x_2, \ldots, x_n, \ldots$ be a realization of a univariate process $(X_n)_{n \in \mathbb{N}}$ sampled over an equispaced time grid $\{t_1, t_2, \ldots, t_n, \ldots\}$ such that

$$X_n = \mu(t_n) + \sigma_{\mathrm{IC}}\varepsilon_n \text{ for } n \in \mathbb{N} \text{ with } \mu(t_n) = \mathbb{E}[X_n], \quad \mathbf{Var}[X_n] \equiv \sigma_{\mathrm{IC}}^2$$

where $\varepsilon_n \overset{\text{i.i.d.}}{\sim} \mathcal{N}(0,1)$. Let $\hat{\mu}_n(t)$ denote the TS estimator proposed in [21] applied to the dataset $(x_1, \ldots, x_n)$. Choosing $\alpha = \frac{3}{5}$ as recommended in [21], the TS signal statistic at the nth period reads as

$$\mathrm{TS}_n = \frac{n^{3/5}}{\sigma_0}\left(\left|\hat{\mu}_n(\tfrac{1}{n}) - \mu_0\right| + \sum_{k=2}^{n}\left|\hat{\mu}_n(\tfrac{k}{n}) - \hat{\mu}_n(\tfrac{k-1}{n})\right|\right). \tag{10}$$

The heuristic behind the choice $\alpha = \frac{3}{5}$ is that it appears to produce a reasonable trade-off between chart performance and right-skewness of the IC run length distribution. The control limits for a selection of reference IC ARL values are given in Table 1. A `Matlab`® implementation of the TS chart is available at https://github.com/mpokojovy/TS.chart.1D.

5   Example

Following the onset of the COVID-19 pandemic late 2019–early 2020, the situation started to get increasingly dramatic early-mid February 2020 [1,18]. On February 3rd, 2020, the US government declared a public emergency due to the COVID-19 outbreak. By February 10th, 2020, the number of COVID-19 deaths in China surpassed that of SARS in 2002–2004. Given the repercussions of the pandemic on the world economy, a crash on the stock market became imminent.

Fig. 1. Meta Platforms, Inc. (FB) daily closing prices. Source: https://www.marketwatch.com/investing/stock/fb

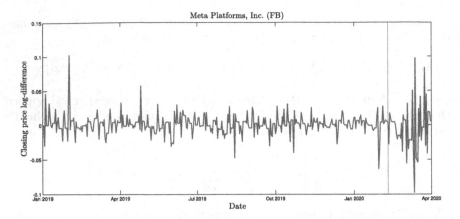

Fig. 2. Daily closing price log-differences.

To illustrate how the proposed methodology could have been applied for FIR financial surveillance right before the crash of Spring 2020, we decided to analyze the closing prices of the Facebook (FB) stock (Meta Platforms, Inc.) displayed in Fig. 1. Assume the analyst began to expect an imminent crash early-mid February. For simplicity, let us assume the monitoring began on February 10, 2020 – the second Monday of February 2020. Suppose the historic FB stock price data from January 1st, 2019 (Wednesday) till February 7th, 2020 (Friday) were used for calibration purposes. To account for missing data due to weekend and holiday market closures, closing prices were linearly extrapolated on such dates.

Adopting the geometric random walk model with log-Student's t increments in Eq. (4), we formed the log-differences $x(t_k) = \log\big(S(t_k)\big) - \log\big(S(t_{k-1})\big)$ displayed in Fig. 2. The vertical line marks the log-difference on the first day of monitoring February 10th, 2020. The historic log-differences before that date (i.e., between January 1st, 2019 and February 9th, 2020) were used to calibrate our geometric random walk model with log-Student's t increments.

Figure 3 plots p-values from a one sample Student's t Kolmogorov-Smirnov test vs. degrees of freedom $\nu > 0$. The global maximum is attained at $\hat{\nu} \approx 2.7$.

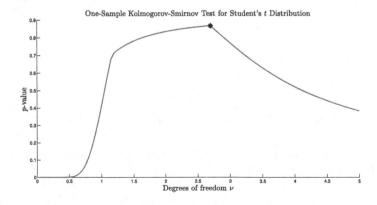

Fig. 3. p-values from a one sample Student's t Kolmogorov-Smirnov test.

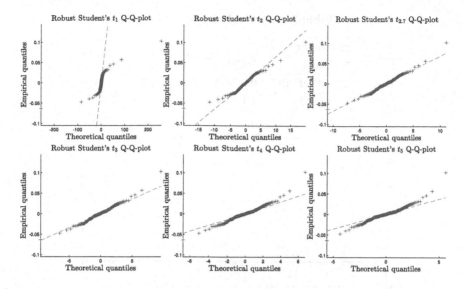

Fig. 4. Robust Student's t Q-Q plots.

This conclusion is consistent with robust Student's Q-Q plots in Fig. 4 for various degrees of freedom ($\nu = 1.0, 2.0, 2.7, 3.0, 4.0, 5.0$). Respective intercept and slope are calculated using robust location and scale estimates

$$\widetilde{\text{median}}\big(\{x(t_k)\,|\,k = 1, 2, \ldots, n_{\text{train}}\}\big) \text{ and } \frac{\widehat{\text{IQR}}\big(\{x(t_k)\,|\,k = 1, 2, \ldots, n_{\text{train}}\}\big)}{t_{\nu,0.75} - t_{\nu,0.75}},$$

where $n_{\text{train}} = 402$ days. Visual inspection confirms the best fit is produced by $\hat{\nu} = 2.7$. Thus, a (rather heavy-tailed) Student's t distribution with $\hat{\nu} = 2.7$ degrees of freedom was selected in lieu of the usual Gaussian distribution to model the log-increments for our time series data.

The log-differences were then robustly gaussianized via

$$y(t_k) = F_{t_\infty}^{-1}\Bigg\{F_{t_{\hat{\nu}}}\Bigg(\frac{t_{\hat{\nu},0.75} - t_{\hat{\nu},0.25}}{\widehat{\text{IQR}}\big(\{x(t_k)\,|\,k = 1, 2, \ldots, n_{\text{train}}\}\big)}\\
\times \big(x(t_k) - \widetilde{\text{median}}\big(\{x(t_k)\,|\,k = 1, 2, \ldots, n_{\text{train}}\}\big)\big)\Bigg)\Bigg\}$$

(11)

with $F_{t_{\hat{\nu}}}(\cdot)$ denoting the cdf of a $t_{\nu=2.7}$-random variable. (Recall $F_{t_\infty}(\cdot)$ is the Gaussian cdf.) To verify the gaussianized log-differences are i.i.d., i.e., the lag size of 0 was adequate for our time series, we performed an ACF plot for the historic gaussianized log-increments in the training set. Using asymptotic 99%-confidence bands, Fig. 5 does not indicate any statistically significant autocorrelation beyond lag 0.

Table 2. Estimated IC standards.

Estimator			
	Usual	MCD	MDPD
$\hat{\mu}$	$3.4169 \cdot 10^{-6}$	0.0095222	-0.0019752
$\hat{\sigma}$	1.0617	1.1429	1.062

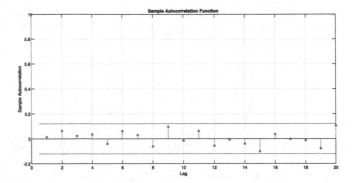

Fig. 5. Autocorrelation (ACF) plot for gaussianized training data log-differences.

Using the gaussianized training data, the in-control (IC) standards μ_{IC} and σ_{IC} were estimated with three sets of estimators: the usual sample mean $\bar{y}$ and standard deviation $\sqrt{s_y^2}$, the MCD estimator pair $\hat{\mu}_{y,\mathrm{MCD}}$ and $\hat{\sigma}_{y,\mathrm{MCD}}$ with the nominal breakdown point of 10% and the MDPD estimator pair $\hat{\mu}_{y,\mathrm{MDPD}}$ and $\hat{\sigma}_{y,\mathrm{MDPD}}$ with $\alpha = 0.1183$ [14] selected to asymptotically match the MCD's breakdown point of 10%. Estimated IC parameters are given in Table 2.

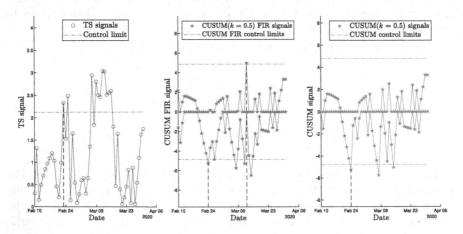

Fig. 6. Process monitoring using $\bar{y}$ and $\sqrt{s_y^2}$ as IC standards.

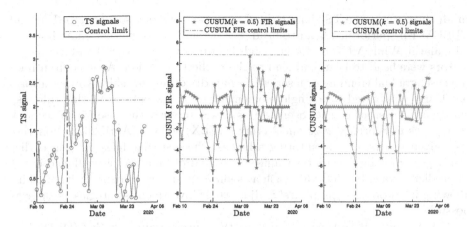

Fig. 7. Process monitoring using $\hat{\mu}_{y,\text{MCD}}$ and $\hat{\sigma}_{y,\text{MCD}}$ as IC standards.

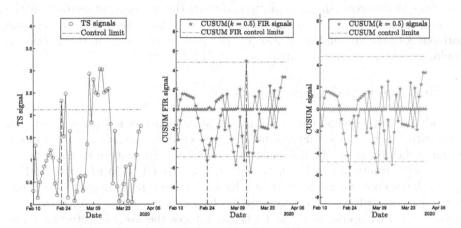

Fig. 8. Process monitoring using $\hat{\mu}_{y,\text{MDPD}}$ and $\hat{\sigma}_{y,\text{MDPD}}$ as IC standards.

Table 3. Crash detection dates.

	Estimator		
	Usual	MCD	MDPD
TS	2/23	2/24	2/23
CUSUM	2/24	2/25	2/24
CUSUM FIR	2/24	2/25	2/24

Beginning process monitoring on Monday, February 10th, 2020, the sequentially observed log-differences were gaussianized using Eq. (11) based on the original location and scale estimators from the training data and subsequently standardized using the IC standards to insure the IC distribution is (approximately) standard Gaussian. Selecting an IC ARL 370 (i.e., an annual average of about one false alarm), the results are displayed in Figs. 6, 7 and 8, respectively.

In all three cases, the TS chart flagged the upset one day sooner. CUSUM and CUSUM FIR performed head-to-head. See Table 3. Also, neither chart issued a false alarm. While MCD performed slightly suboptimally, MDPD and usual estimators were head-to-head and lead to the earliest detection. As it is oftentimes the case, gradual trending or a series of small discontinuous shifts cannot always be clearly pinpointed to a single moment of time. Exceptions to this rule also exist when the stock crashes overnight as it was recently the case with Facebook (FB) stock in February 2022 and Netflix (NFLX) stock in April 2022. Nonetheless, given two process monitoring schemes, the one that flags a crash earlier can *a posteriori* be declared a winner. The goal of financial surveillance is not to predict a crash but flag the crash as soon as it starts to manifest itself, thus, allowing investors to appropriately adjust their investment strategies as early as possible.

Using robust IC estimates (such as the ones obtained with MDPD) is a prudent statistical practice as it alleviates potential masking effects (which did not seem to be severe in our situation) due to the presence of outliers in the training dataset, e.g., due to (micro-)crashes or rallies over the training time horizon. A Matlab® implementation of the analyses presented in this Section is available at https://github.com/mpokojovy/TS.FIR.finsurv.

6   Summary and Conclusions

To address the problems posed by financial surveillance of a geometric random walk with log-Student's t increments, we proposed a sequential process monitoring technique using the TS chart [21] with the FIR feature to consistently detect early structural break in financial data. Unlike CUSUM FIR, the TS FIR approach does not require prior knowledge about the type or magnitude of shift in the process drift which categorizes the latter as a reference-free technique. In turn, the FIR property of the TS chart reduces the average detection time. An example involving an application of the proposed methodology to real-world financial data from Meta Platforms, Inc. (FB) daily stock closing prices for early detection of the February 2020 crash. The IC standards were estimated with the usual non-robust approach and robust MCD and MDPD estimators. The results were compared across the TS, CUSUM and CUSUM FIR charts. Based on the empirical results presented, the TS chart of [21] was demonstrated to offer advantage of the two prominent competitors. This is consistent with the extensive simulation results reported in [21]. A set of Matlab® codes to reproduce the example from Sect. 5 is provided online.

References

1. A timeline of COVID-19 developments in 2020. AJMC (2021). https://www.ajmc.com/view/a-timeline-of-covid19-developments-in-2020 Accessed 7 Mar 2022
2. Anum, A.T., Pokojovy, M.: A hybrid method for density power divergence minimization with application to robust univariate location and scale estimation (under revision) (2022)

3. Astill, S., Harvey, D.I., Leybourne, S.J., Taylor, A.M., Zu, Y.: CUSUM-based monitoring for explosive episodes in financial data in the presence of time-varying volatility. J. Financ. Econometrics, 1–41 (2021)
4. Basu, A., Harris, I.R., Hjort, N.L., Jones, M.C.: Robust and efficient estimation by minimising a density power divergence. Biometrika **85**(3), 549–559 (1998)
5. Beccar-Varela, M.P., Mariani, M.C., Tweneboah, O.K., Florescu, I.: Analysis of the Lehman brothers collapse and the flash crash event by applying wavelets methodologies. Phys. A **474**, 162–171 (2017)
6. Blondell, D., Hoang, P., Powell, J.G., Shi, J.: Detection of financial time series turning points: a new CUSUM approach applied to IPO cycles. Rev. Quant. Financ. Acc. **18**(3), 293–315 (2002)
7. Brigo, D., Dalessandro, A., Neugebauer, M., Triki, F.: A stochastic processes toolkit for risk management: mean reverting processes and jumps. J. Risk Manag. Fina. Inst. **3**(1), 65–83 (2009)
8. Carlisle, M., Hadjiliadis, O., Stamos, I.: Trends and Trades. Handbook of High-Frequency Trading and Modeling in Finance, pp. 1–49. Wiley, Hoboken (2016)
9. Frisén, M.: Optimal sequential surveillance for finance, public health, and other areas. Seq. Anal. **28**(3), 310–337 (2009)
10. Han, D., Tsung, F.: A reference-free CUSCORE chart for dynamic mean change detection and a unified framework for charting performance comparison. J. Am. Stat. Assoc. **101**(473), 368–386 (2006)
11. Hubert, M., Debruyne, M.: Minimum covariance determinant. Wiley Interdisc. Rev. Computat. Stat. **2**(1), 36–43 (2010)
12. Lorden, G.: Procedures for reacting to a change in distribution. Ann. Math. Stat. **42**, 1897–1908 (1971)
13. Lucas, J.M., Crosier, R.B.: Robust CUSUM: a robustness study for CUSUM quality control schemes. Commun. Stat. - Theory Methods **11**(23), 2669–2687 (1982)
14. Lucas, J.M., Crosier, R.B.: Fast initial response for CUSUM quality-control schemes: give your CUSUM a head start. Technometrics **42**(1), 102–107 (2000)
15. Mariani, M.C., Bhuiyan, M.A.M., Tweneboah, O.K., Huizar, H.G.: Forecasting the volatility of geophysical time series with stochastic volatility models. Int. J. Math. Comput. Sci. **11**(10), 444–450 (2017)
16. Mariani, M.C., Tweneboah, O.K.: Modeling high frequency stock market data by using stochastic models. Stoch. Anal. Appl. **40**, 1–16 (2021)
17. Moustakides, G.V.: Optimal stopping times for detecting changes in distributions. Ann. Stat. **14**(4), 1379–1387 (1986)
18. A timeline of the coronavirus pandemic. The New York Times (2021). https://www.nytimes.com/article/coronavirus-timeline.html. Accessed 7 Mar 2022
19. Omelchenko, V.: Parameter estimation of sub-Gaussian stable distributions. Kybernetika **50**(6), 929–949 (2014)
20. Pepelyshev, A., Polunchenko, A.S.: Real-time financial surveillance via quickest change-point detection methods. Stat. Interface **10**(1), 93–106 (2017)
21. Pokojovy, M., Jobe, J.M.: Univariate fast initial response statistical process control with taut strings. J. Appl. Stat. **49**, 1–23 (2021)
22. Prestele, C.: Credit portfolio modelling with elliptically contoured distributions-approximation, pricing, dynamisation (Doctoral dissertation, Universität Ulm) (2007)
23. Rousseeuw, P.J., Leroy, A.M.: Robust Regression and Outlier Detection. Wiley, Hoboken (1987)

Evaluation of High-Value Patents in the Reverse Innovation: A Theory and Case Study from China

Chen Hui and Wei Tie[✉]

School of Business, Guangxi University, Nanning 530004, China
weitie@gxu.edu.cn

Abstract. The connotation and value evaluation of high-value patents are context-dependent. In the context of reverse innovation of Chinese local enterprises, the paper discusses the connotation and value evaluation methods of high-value patents. The paper proposes a theoretical framework of high-value patents in the context of reverse innovation of local enterprises, and analyzes the connotation and value characteristics of high-value patents from three aspects of market control, legal stability and technological advancement. It uses the natural Data mining methods to evaluate the value of high-value patents in this context and takes BYD as an example to analyze and evaluate the high-value patents in its reverse innovation, and discusses the characteristics of its value changes. The research provides new theory and methodology to identify the connotation and characteristics of high-value patents in reverse innovation of local enterprises in developing countries, and assess their values as well.

Keywords: Reverse innovation · High-value patent · Evaluation of patent

1 Introduction

Since the beginning of the 21st century, a phenomenon that an innovative product or technology originating from a developing country or emerging market spreads back to developed countries has attracted theoretical and practical attention. For example, in 2002, the Chinese team of General Electric Company developed a portable ultrasonic device which was very successful in the Chinese market. After that, this product originally designed for the backward market was also unexpectedly popular in the United States. This is different from the traditional innovation model that originated in developed countries and then spread to developing countries. The direction of innovation diffusion is "reverse", so it is known as "Reverse Innovation" [1].

In recent years, with the rise of local companies in developing countries, reverse innovation has increasingly occurred in local companies of developing countries, such as BYD's battery in China and Tata Motors in India [2, 3]. Although the number of overseas patent applications of Chinese companies has increased significantly in recent

H. Han and E. Baker (Eds.): SDSC 2022, CCIS 1725, pp. 34–48, 2022.
https://doi.org/10.1007/978-3-031-23387-6_3

years, the overall quality of overseas patents of Chinese companies is not high, and there is still a large gap in the core patent layout compared with developed countries [4, 5]. This has also led to many local companies experiencing repeated setbacks in reverse innovation, especially when new products or new technologies are traced back to the markets of developed countries.

In practice, China has generated a strong need for the cultivation of high-value patents from the government to the enterprise [6]. However, the current researches still have different understandings of the concept of high-value patents, it also shows that the concept of high-value patents is highly context dependent. Therefore, this paper intends to explore the connotation and characteristics of high-value patents in the context of reverse innovation by local companies, and discuss the evaluation methods of their value based on the connotation characteristics, so as to provide theoretical and decision-making references for the cultivation of high-value patents in reverse innovation by local companies. The main contributions of this paper are:

(1) constructs a theoretical framework of high-value patents in the context of reverse innovation of local enterprises, proposes the connotation and characteristics of high-value patents in the context of reverse innovation of local enterprises, and provides a new research perspective to promote the formation of high-value patent theories.
(2) proposes a new method of value evaluation of patents, that is, based on an open patent document database, using natural language processing and other data mining methods, objectively analyze and evaluate the high-value patents in the reverse innovation of local enterprises.

2 Literature Review and Theoretical Foundation

2.1 Reverse Innovation of Local Companies

Reverse innovation is not only a strategy for multinational corporations to innovate and internationalize, but also an inevitable evolution of the innovation model of local enterprises in developing countries [7]. After the local enterprises accumulate certain innovation ability, their innovation mode gradually changes from traditional imitation innovation and low-cost innovation to independent development mainly reverse innovation [8]. At present, the research based on the perspective of developing countries is mostly carried out around the case of local enterprises, they discussed the conceptual types and influencing factors of reverse innovation based on typical cases of successful reverse innovation of local enterprises [9–13].

In addition, the research on reverse innovation basically focuses on the impact of reverse innovation on corporate behavior [14]. Reverse innovation has a positive impact on social effects such as local citizen behavior in developing countries and these effects can promote the improvement of local enterprises' innovation capabilities [15].

However, there is still a lack of in-depth discussion on how local enterprises form intellectual property rights with control advantages in reverse innovation [16–18].

2.2 High-Value Patents

In recent years, with the rapid growth of patent applications in China, the quality and value of patents have attracted more and more attention from the whole society. The evaluation of patent value has received unprecedented attention, and high-quality, high-value patents have gradually become the basis for companies to maintain their core competitiveness. However, the concept of high-value patents has not yet been uniformly defined. High-value patents are different from traditional core patents and patent quality, traditional core patents ignore the market value of patents, while high-value patents are a comprehensive concept. Its value is mainly reflected in the high technical content of patents, high legal stability and high market value [6, 19–21]. Although there is no uniform definition of high-value patents, these researches believe that high-value patent is a strategic concept and brings strong market control to competitors, and its value is mainly reflected in three dimensions: high-level technical characteristics [6], legal claims with high stability [20, 21], and market control with a high strategic position [22, 23].

2.3 Evaluation of Patent Value

The existing patent value evaluation research mainly focuses on the construction of the evaluation index system and the selection of methods [24–26]. At present, there is a relatively mature patent value evaluation system, and there are many types of evaluation indicators for patent value, involving technology, economy, law, policy, organization, etc. In terms of evaluation methods, the existing research generally follows the three basic methods of asset evaluation: income method, cost method and market method, as well as other methods derived and improved on this basis, such as real option method [27], fuzzy evaluation method [28], neural network method [29] and reference network method [30, 31], etc., but these methods and evaluation indicators have defects are difficult to quantify.

At present, the evaluation of high-value patents mainly used the existing methods [32, 33]. As the concept of high-value patents has strong situational dependence, the value evaluation of high-value patents also needs to be analyzed from its connotation and characteristics in a specific context. There is still a lack of research on the connotation and value evaluation of high-value patents combined with the reverse innovation situation of local enterprises.

2.4 The Connotation of High-Value Patent in the Reverse Innovation

In the reverse innovation of local enterprises, high-value patents have specific connotations and value characteristics. First of all, in terms of market control, the high-value patents in the reverse innovation of local enterprises should be reflected in the promotion of the new products or new technologies of local enterprises to retroactively trace back to the markets of developed countries. In the fierce domestic and overseas market development, it plays a role of escort and gaining competitive advantage. Secondly, for the high-value patents in the reverse innovation of local enterprises to play the role of market control, they must have certain technological leadership and strong legal stability.

Based on this, this paper defines the connotation of high-value patents in reverse innovation of local companies as follows: the high-value patents in the reverse innovation of local enterprises refer to the patents which come from the local enterprises and can truly promote the success of their new products or new technologies in the local market and then retrospectively return to the developed country market. In the fierce domestic and overseas market development, high-value patents play an important role of protecting technologies and gaining competitive advantage. They significantly show the value in three aspects, i.e., market control, technological advancement and legal stability, shown in Fig. 1.

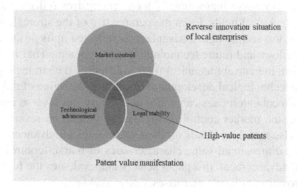

Fig. 1. The value embodiment of high-value patents in the reverse innovation.

3   The Methods of Selection and Evaluation of High-Value Patents

3.1   The Selection of High-Value Patents in the Reverse Innovation

First of all, according to the previous analysis of the connotation and value characteristics of high-value patents in reverse innovation of local enterprises, this paper believes that in reverse innovation of local enterprises, its high-value patents must be closely related to its successful products (or technologies) of reverse innovation. Because the reverse innovation products of local enterprises not only achieve success in the domestic market, but also successfully explore the market of developed countries, and the patents closely related to their products also reflect their high market control and technological advantages. Therefore, firstly, patents closely related to successful reverse innovation products of local enterprises are selected as high-value patents preliminarily shortlisted. Among them, close correlation refers to that the technical field covered by the patent is the core technology of the reverse innovation product.

Secondly, further select the patents with strong legal stability among these patents Studies have pointed out that the difference in the number of patent rights requirements can reflect the legal stability of patents to a large extent. It is generally believed that the greater the number of claims in a patent, the greater the scope of protection of the rights, and the stronger the legal stability of the patent [34–36]. Therefore, the value of a patent in the legal dimension can be characterized by the number of claims, and the greater the number of claims, the higher the legal stability of the patent. This paper will further screen out the number of legal claims above the average level of related fields as high-value patents.

After the above two steps of screening, the selected patents already possess the basic characteristics of market control and legal stability of the high-value patents in the reverse innovation of local enterprises (as shown in Fig. 2). At this time, the value of high-value patents is mainly reflected in the advanced nature of its technology. This is also the main basis for this paper to evaluate the high-value patents in the reverse innovation of local enterprises. As mentioned above, the technological advancement of high-value patents in reverse innovation of local enterprises reflects their advantages in certain specific technical fields. From the perspective of the strategic significance of high-value patents, this technological advantage is reflected in its position in the industry's current technology and future technological development. That is, the foundation and core position in the current technical field, as well as the strategic position occupied by the future technological development trend. It is the value of this technological advancement that local enterprises can continue to carry out reverse innovation in the end. In this sense, the market control of high-value patents in reverse innovation of local enterprises also mainly depends on their technological advancement. Therefore, based on the three-dimensional value characteristics of market control, legal stability and technological advancement, this paper selects and evaluates the high-value patents in the reverse innovation of local enterprises.

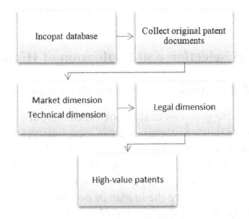

Fig. 2. The selection process of high-value patents.

3.2   The Process of Evaluation of High-Value Patents

Based on the above analysis, this paper collects the "high-value" characteristics of high-value patents of local enterprises in three dimensions of law, market and technology, and proposes a three-dimension framework of the value characteristics of high-value patents.

Under this framework, high-value patents are screened and evaluated in the context of reverse innovation of local enterprises. After being screened by the characteristics of market control and legal stability, the value of high-value patents is reflected in technological advantages. That is, the foundation and core position in the current technical field, as well as the strategic position occupied in the future technological development trend. Therefore, we evaluate the value of high-value patents in the reverse innovation of local enterprises through the analysis of market control and legal stability, and constructs a value indicator that can reflect its technological advancement to quantify and comprehensively evaluate its value.

The Technology Centrality Index (TCI) is based on the patent network analysis model, which reflects the foundation and core position of the target patent in the field. Based on the context of reverse innovation by local companies, this paper proposes the use of Technology Centrality Index (TCI) to comprehensively measure the value of high-value patents. We use Python to conduct data mining and text processing on patent data. The calculation steps of the TCI index constructed in this paper are as follows:

Step 1: Establish a patent data vocabulary in key technical fields closely related to the reverse innovation products of local enterprises.

Step 2: Calculate the word frequency matrix of the patent data based on the patent thesaurus, and integrate the data into a keyword vector. The keyword vectors from patent 1 to patent m are as follows:

$$
\begin{aligned}
&\text{Patent 1}: \quad (k_{11}, k_{12}, k_{13}, \ldots, k_{1n}) \\
&\text{Patent 2}: \quad (k_{21}, k_{22}, k_{23}, \ldots, k_{2n}) \\
&\qquad\vdots \qquad\qquad\qquad \vdots \\
&\text{Patent m}: \quad (k_{m1}, k_{m2}, k_{m3}, \ldots, k_{mn})
\end{aligned}
$$

For example, in the document of Patent 1, as described above, the first keyword appears k_{11} times, the second keyword appears k_{12} times, and so on.

Step 3: Reduce the dimensionality, perform PCA processing on the matrix generated in the second step, and select the principal components with a contribution greater than 80% to form a new principal component matrix.

Step 4: The principal component matrix is screened, and the principal components that do not belong to the key technical fields are eliminated, and the retained principal components form a new principal component matrix.

Step 5: According to the new principal component matrix, the Euclidean distance formula is used to calculate the distance between patents and establish the relationship between patents. If the principal component vectors of patent i and patent j are respectively defined as $(k_{i1}, k_{i2}, \ldots, k_{in})$ and $(k_{j1}, k_{j2}, \ldots, k_{jn})$, the Euclidean distance between the two vectors is calculated as follows:

$$E_{ij}^{d} = \sqrt{(k_{i1} + k_{j1})^2 + (k_{i2} + k_{j2})^2 + \cdots + (k_{in} + k_{jn})^2} \tag{1}$$

Step 6: The Euclidean distance matrix ($E^d{}_{ij}$ matrix) is composed of all the Euclidean distance values between all vectors. However, in order to draw the patent network diagram, the $E^d{}_{ij}$ matrix must be standardized. In this step, the real value of the $E^d{}_{ij}$ matrix needs to be converted into a standardized value, the E^d_{ij} matrix is constructed, and then the dichotomy is performed from 0 to 1.

$$E^s_{ij} = \frac{E^s_{ij}}{Max(E^d_{ij}, i = 1 \cdots m, j = 1 \cdots m)} \tag{2}$$

Step 7: If the critical value p is to be exceeded, the connection between the two patents is considered to be weak, and it is recorded as 0, otherwise it is recorded as 1, and the unit of the E^s_{ij} matrix is converted into a matrix I_{ij} containing only 0 and 1:

$$I_{ij} = \begin{cases} 1, E^s_{ij} < p \\ 0, E^s_{ij} > p \end{cases} \tag{3}$$

The I_{ij} matrix only contains 0 and 1, where if patent i is strongly connected to patent j, I_{ij} is 1, and if patent i is weakly connected to patent j or not connected at all, I_{ij} is 0. The determination of the critical value p is a task of trial and error, and a reasonable cut-off value must be selected so that the network structure becomes clearly visible (the p value is selected as 0.5 in this paper).

Step 8: Calculation of TCI. TCI is an index for judging the location of the target patent in the overall network. Through the above steps, this article uses it to finally comprehensively evaluate the value of high-value patents. Its calculation formula is:

$$TCI = \frac{C_i}{n-1}, C_i = \sum_{i=1}^{j=1 \cdots n} I_{ij} \tag{4}$$

where n represents the number of patents. It measures the relative importance of the target patent by calculating the correlation density between the target patent and other patents. C_i is the connection between each patent and other patents. The higher the TCI, the greater the impact on other patents, which means that the target patent is at the core and occupies a leading position in the technical dimension.

The technology centrality index in the patent network model can be interpreted as the connection between the target patent and other patents. Therefore, the higher the technology centrality index, the greater the impact on other patents. The technology centrality index takes the patent document as the initial input, conducts text analysis on it, and integrates it into a keyword vector. Through the technical processing of the vector matrix, with the aid of the patent network analysis model, objectively calculate the position of the target patent in the patent network, and then evaluate the value of the target patent. Compared with the previous high-value patent evaluation methods, the biggest advantage of patent network model and technology centrality index lies in its testability and objectivity, which can realize the evaluation of high-value patents in different fields. To a certain extent, it makes up for the subjective shortcomings of evaluating patent value by constructing a multi-dimensional index system in the past, and solves the disadvantages that are difficult to quantify in the three evaluation methods of income method, cost method and market method. Furthermore, the technology centrality index is quite different from the traditional citation analysis. The relationship between the two patents in the technology centrality index is characterized by a keyword vector. It explains the internal structure between patents and reflects the value of patent quality.

4   Case Study

4.1   BYD Company and Its Reverse Innovation

BYD is a local manufactural company of China and was founded in 1995. After more than 20 years of fast growth, BYD has established over 30 industrial parks and its products involve four major industries: IT, automotive, new energy and rail transit [37, 38]. Since BYD launched the F3 model successfully in 2005 at home and abroad, more and more BYD products have successfully entered the markets of developed countries [39]. The pure electric buses developed by BYD have spread to more than 50 countries including the United States, the United Kingdom, and Australia, and have become typical products of successful reverse innovation [39]. It can be seen that BYD's reverse innovation products are mainly pure electric buses. Pure electric bus refers to the new energy vehicle with power battery as the only energy source. A pure electric bus is mainly composed of an electric drive control system, a chassis, a body and various auxiliary equipment [38]. The specific structure is shown in Fig. 3.

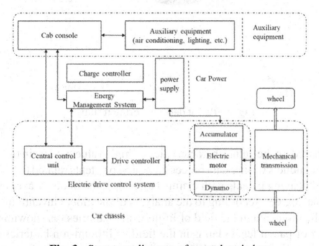

Fig. 3. Structure diagram of pure electric bus.

Batteries, motors, and electronic control (three-electric technology) are the hotspots of new energy vehicle technology. BYD's mature three-electric technology gives BYD a unique advantage in the development and production of new energy vehicles. As a leading company in China's new energy vehicles, BYD adheres to an independent and controllable technological innovation strategy, while mastering core technologies such as power batteries and IGBTs. In the battery field, the highly safe lithium battery developed by BYD has solved the current global problems in the safety and battery life of electric vehicle batteries [38]. The success of BYD pure electric bus indicates the benchmarking position of BYD enterprise in power battery technology and quality, intelligent manufacturing, promotion and application. As the heart of pure electric bus, battery costs account for about 30 to 40% of the cost of the entire vehicle [38].

4.2 Selection of High-Value Patents in BYD'S Reverse Innovation

First, we search the patents of BYD from the INCOPAT database and obtain a total of 5927 patents as the sample. As mentioned above, the initial shortlisted patents are selected based on the value characteristics of high-value patents with strong market control. The high-value patents in BYD's reverse innovation must be closely related to the products (or technologies) that succeeded in reverse innovation. That is to choose a patent closely related to BYD's lithium-ion battery technology for pure electric buses. In this paper, cluster analysis (as shown in Fig. 4) is used to select 426 patents of BYD company in the field of lithium-ion battery as the research object.

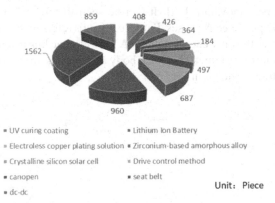

Fig. 4. Patent data clustering results.

Secondly, based on the value characteristics of high-value patents in the legal dimension, this paper conducts further data processing on 426 patents related to BYD's lithium-ion battery field, leaving patents with claims larger than the industry average and calculating their value degree. According to the analysis of the global invention authorization and patent application data in the field of lithium-ion batteries (as shown in Fig. 5), the average number of patent legal claims in the field of lithium-ion batteries is 10. Based on this, 187 high-value patents in BYD's reverse innovation were selected.

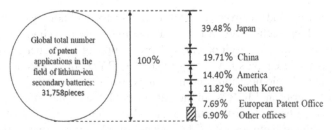

Fig. 5. Distribution of patents in different countries.

After the above two steps of filtering, as the research object of the article 187 of the patent has the characteristics of market control and stability of law, the company at this time of the high value of reverse innovation patent value embodies in technical advantage, and the technology advantage is the company continued competitiveness in domestic and foreign markets and the basis of market control. Therefore, 187 patents of BYD company in the field of lithium battery are finally determined as high-value patents in reverse innovation. This paper analyzes the strategic position of BYD's patent in the field of lithium battery by constructing patent network analysis model and using technology centrality index.

4.3   Discussion

The calculation results of value degree are shown in Fig. 6.

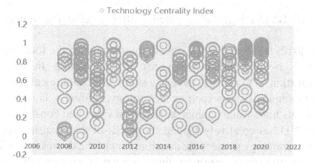

Fig. 6. The value of BYD's high-value patents.

The results show that the value degree of high value patents of BYD company has obvious difference. Among them, CN109428053B is the patent with the highest value of 0.98. Specifically, the patent relates to a lithium battery cathode plate and its preparation method and all-solid-state lithium battery and pre-solid-state lithium battery. The minimum value of BYD's high-value patent is 0.02 (CN101783422B), which was announced in 2012. From 2008 to 2012, BYD's main exporting countries were developing countries, and its advantages in developed markets were not obvious. BYD's value was also low due to its early technological catch-up. In addition, the average value of its high-value patents is 0.61. From the figure, the distribution of BYD's high-value patents gradually concentrates on the average.

The average value of BYD's high-value patents is 0.61, and the value of high-value patents at different stages of reverse innovation has significant differences. Meanwhile, in order to better discover the time distribution of the value of these high-value patents, we conduct a comparative analysis of the changes in the value of these patents by year, as shown in Fig. 7.

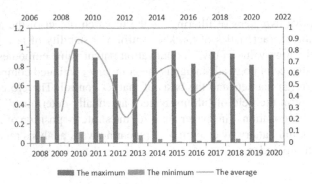

Fig. 7. Value distribution of BYD's high-value patents.

From the Fig. 7, we find:

(1) The value of BYD's high-value patents after reverse innovation has steadily increased.

   Before 2012, BYD was in the development stage of the local market, and the value of its high-value patents was constantly fluctuating. In the early stage of reverse innovation, BYD was in the early stage of technological catch-up, affected by multiple complex factors such as technology, economy, law, policy, and organization, and its high-value patents were low in value and constantly changing. At this stage, BYD has creatively tapped the needs of the local market and used its own battery technology to meet the development of the domestic market, gradually forming a monopolistic situation, gaining market development effects and sustained competitive advantages. After 2012, BYD's products and technologies have been traced back to developed countries. In order to adapt to the fiercely competitive overseas market, BYD's core technology has been continuously improved, breaking through technical barriers in foreign markets and gaining continuous corporate competitiveness. Therefore, in the developed market stage (after 2012), THE value of BYD's high-value patents has steadily increased.

(2) High-value patents are the basis for BYD to maintain its core competitiveness.

   In the local market development stage, the number of high-value patents of BYD enterprises is small, and the competition situation in the domestic market is relatively relaxed. With its own leading technology in the battery field, BYD took the lead in opening the domestic new energy vehicle market and became the leader in the new energy field in the domestic market. In the market stage of developed countries, BYD enterprises rely on overseas markets, reduce competition costs through continuous technological innovation, and achieve technological catch-up. At this stage, BYD's high-value patents are still the foundation for the enterprise to maintain its core competitiveness.

(3) The high value patents of BYD company have significant differences in different stages of reverse innovation.

   In the development stage of the local market, the quality of BYD's high-value patents is more polarized, and the minimum value rises rapidly after the reverse to the developed countries. The reason is that BYD enterprises are more likely to open the domestic new energy market as market followers when developing the local market, and constantly imitate foreign developed enterprises in terms of technology. Therefore, its immature technology leads to low value of high-value patents. When it comes to the market of developed countries, BYD is no longer simply imitating and following, but constantly making up for its shortcomings in technology and improving its research and development and management capabilities. Therefore, the value degree of its high-value patents is improved, and the minimum value is also developed with the success of reverse innovation. The development of high-value patents promotes the improvement of BYD's reverse innovation ability, and enables BYD to establish a foothold in overseas markets, thus continuously expanding market share.

Based on the above analysis, the method of evaluating high-value patents based on the patent network analysis model proposed in this paper using technology centrality index can not only evaluate the high-value patents of enterprises in a certain field, but also on this basis, to conduct a more in-depth analysis of the evaluation results of high-value patents, excavate high value at different stages of reverse innovation patent technology value variation characteristics, formation and local companies reverse innovation practice closely combining the basic theory of the high value of patent framework and assessment methods.

5 Conclusions

This paper puts the research on high-value patents in the context of reverse innovation of local companies, and constructs a theoretical analysis framework for the connotation and characteristics of high-value patents in this context. On this basis, a new method of patent valuation is proposed, and the case of BYD is analyzed. The main conclusions are as follows:

(1) The connotation of high-value patents in reverse innovation of local enterprises is reflected in the strategic patents that can promote the products (or technologies) of local enterprises to go back to developed countries and obtain sustainable competitiveness for enterprises in the fierce overseas market through continuous technological catch-up and innovation. The high-value patents in the reverse innovation of local enterprises have the unified value characteristics of market control, technological advancement and legal stability. These characteristics are the high-value patents that can protect the trend of reverse innovation of local enterprises and the key to sustained success.

(2) Based on the connotation and characteristics of high-value patents in reverse innovation of local enterprises, this paper proposes a patent value evaluation method based on original patent documents, using natural language processing and data mining for text processing. This method makes up for the defect of evaluating patent value by constructing multi-dimensional index evaluation system, and has good objectivity, accuracy and practicability. It can realize high-value patent valuation in reverse innovation of local enterprises in different fields.

(3) Through the evaluation of BYD's high-value patents, it is found that BYD's high-value patents have different manifestations in different periods of reverse innovation. The polarization of the value of BYD's high-value patents has improved as BYD's products (or technologies) are traced back to the markets of developed countries, and the value of its high-value patents has tended to rise steadily. By relying on the technical barriers formed by high-value patents, BYD took the lead in opening up the domestic new energy market, constantly making up for shortcomings, quickly entering overseas markets, and finally achieving technological catch-up. Therefore, high-value patents play an indispensable role in improving the ability of local enterprises to reverse innovation. Through high-value patents, enterprises build their own technical barriers and form core competitiveness, thus escorting the development of reverse innovation and enabling local enterprises to obtain a sustainable and stable development trend in overseas markets.

In the reverse innovation of local enterprises, the value of high-value patents continues to dynamically evolve as the stage of reverse innovation changes. This paper only compares the value degree of high-value patents in different stages of reverse innovation, but still lacks the dynamic evolution analysis of high-value patents. Therefore, follow-up research can analyze the dynamic evolution characteristics and influencing factors of high-value patents more deeply, and reveal the mechanism of the evolution of high-value patents at a deeper level.

Acknowledgments. This research is supported by the National Natural Science Foundation of China (No. 72062002).

References

1. Baozhi, W., Lin, L., Gang, R.: Take the advantage of examination resources and promote the foster of hight-quality patent. China Invent. Patent **15**(4), 25–29 (2018)
2. Boxin, L., Xiangfeng, H., Ning, L.: Dose IPR strength the OFDI reverse innovation spillover in China? China Soft Science **3**, 46–60 (2019)
3. Chaohui, C., Zhijuan, Z.: An value assessment model of patent portfolio and empirical analysis based FANP under the circumstances of big data: a case study of DJI. Sci. Technol. Progress Policy **37**(5), 18–26 (2020)
4. Chunpeng, Y., Haihua, W.: Real options to the value of patent right. Syst. Eng. Theory Practice **6**, 101–104 (2002)
5. Fisch, C., Sandner, P., Regner, L.: The value of Chinese patents: an empirical investigation of citation lags. China Econ. Rev. **45**, 22–34 (2017)
6. Govindarajan, V.: A reverse-innovation playbook. Strat. Direct. **28**(9) (2012)
7. Guoxin, L.: BYD Pure Electric Bus Entered Romania for the First Time. People's Public Transport. **2**, 94 (2021)
8. Gupeng, Z., Xingdong, C.: The patent value based on renewal periods: a model reconstruction with the hypothesis of exponentially distributed return. J. Indust. Eng. Eng. Manage. **27**(4), 142–149 (2013)
9. Hongyun, L., Qingpu, Z.: The small technology-based new ventures radical innovation process under the perspective of knowledge management. Sci. Sci. Manage. S.&T **36**(3), 143–151 (2015)

10. Immelt, J.R., Govindarajan, V., Trimble, C.: How GE is disrupting itself. Harv. Bus. Rev. **87**(10), 16–17 (2009)
11. Jianfei, H., Xiang, C.: Brief analysis on identification of core patent and high-value patent. China Invent. Patent **14**(8), 27–31 (2017)
12. Jin, C., Heng, H.: Reverse innovation: a new class of innovation. Sci. Technol. Progress Policy **28**(8), 1–4 (2011)
13. Juan, S., Guowei, D.: Research on the path of reverse innovation for latecomer firms in emerging markets-based on the case study on Huawei. Sci. Technol. Progress Policy **34**(2), 87–93 (2017)
14. Ling, L.: All round BYD. Manager **9**, 20–31 (2020)
15. Meiling, N.: Research on Key Success Factors of Reverse Innovation in Electronic Information Industry. Dalian Polytechnic University (2016)
16. Nana, X., Yusen, X.: Research on the path of reverse innovation of enterprises: based on grounded theory analysis of longitudinal cases. Chinese J. Manage. **12**(11), 1579–1587 (2015)
17. Ping, C., Zeng, Y.: Reverse trends and its framework for research. Sci. Technol. Progress Policy **32**(17), 22–26 (2015)
18. Qingbing, Z.: Pure Electric Bus Body Structure Design and Analysis. Shenyang University of Technology (2018)
19. Qinghai, L., Yang, L., Sizong, W., Xu, X.: Patent value indicators and their structure. Stud. Sci. Sci. **2**, 281–286 (2007)
20. Rui, L., Wei, Z., Junfeng, R., Zhijun, K.: The citation analysis of high value companies patent: a case study of fortune global 500. J. China Soc. Sci. Techn. Inform. **34**(9), 899–911 (2015)
21. Corsi, S.: Reversing the International Flow of Innovation: How Does Chinese Market Trigger Reverse Innovation? SSSUP (2012)
22. Suzuki, J.: Structural modeling of the value of patent. Res. Policy **40**(7), 986–1000 (2011)
23. Tianqi, M., Xing, Z.: Research on the connotation and controlling factors of high value patent. China Invent. Patent **15**(3), 24–28 (2018)
24. Wang, B., Hsieh, C.: Measuring the value of patents with fuzzy multiple criteria decision making: insight into the practices of the Industrial Technology Research Institute. Technol. Forecast. Soc. Chang. **92**, 263–275 (2015)
25. Xiaofei, L.: Go to Japan to explore the advanced road of BYD's "Export Freedom". Commercial Vehicle News **19**, 3 (2019)
26. Xiaojun, H., Jin, C.: An investigation on indicators for patent value based on structured data of patent documents. Stud. Sci. Sci. **32**(3), 343–351 (2014)
27. Xiaoli, W., Xuezhong, Z.: The indicator system and fuzzy comprehensive evaluation of patent value. Sci. Res. Manage. **2**, 185–191 (2008)
28. Xiuchen, H., Yi, L.: Theory and practice analysis of cultivating high value patents. China Invent. Patent **14**(12), 8–14 (2017)
29. Xuefeng, W., Xiaoxuan, L., Donghua, Z.: Research on patent value indicators. Sci. Manage. Res. **26**(06), 115–117 (2008)
30. Xueseng, L.: On the relationship between quantity and quality of chinese patent application. China Invent. Patent **14**(03), 34–38 (2017)
31. Yaodi, L.: Research on key factors of acceleration of internationalization of the latecomer enterprises on the basis of comprehensive advantage. Sci. Sci. Manage. S.& T. **35**(06), 128–136 (2014)
32. Ye, G., Guozhong, C., Shengkun, W.: Research on evaluation method of high value patents based on functional analysis. Sci. Technol. Manage. Res. **40**(23), 187–196 (2020)
33. Ying, D., Jin, C.: Reverse innovation for sustainable development. Technol. Econ. **29**(01), 9–12+102 (2010)

34. Yonggui, W., Na, W.: Does reverse innovation foster subsidiaries' power and MNC local citizenship? An empirical study of subsidiaries of large multinational corporations in China. Manage. World **35**(04), 145–159 (2019)
35. Yun, L., Lu, L., Zhe, Y., Cheng, Y.: Characteristic analysis of the CNT technical innovation based on patentometrics. Sci. Res. Manage. **37**(01), 337–345 (2016)
36. Yunhua, Z., Jing, Z., Yan, L., Yin, X.: Study on evaluation for patent value based on machine learning. Inform. Sci. **31**(12), 15–18 (2013)
37. Yusen, X., Xiangchun, W., Nana, X., Supeng, Z.: A Research on the process and key factor of reverse innovation for latecomer engineering and technical service enterprises. Sci. Res. Manage. **38**(06), 9–17 (2017)
38. Zhengang, Z., Chuanpeng, Y.: Research on the Development of International Cutting-edge Technology. South China University of Technology Press, Guangzhou (2021)
39. Qin, L., Shigan, Y., Youhua, L.: High-value patent evaluation method existing problems and countermeasures. Sci. Technol. Manage. Res. **42**(04), 147–152 (2022)

Research on the Influence of Different Fairness Reference Points on the Supply Chain Enterprises

Hui -min Liu[✉], Hui Hao, and Yameng Zou

School of Business, Guangxi University, Nanning 530004, China
lhmmath@163.com, haohui@st.gxu.edu.cn

Abstract. We consider a supply chain system consisting of a fair-neutral supplier and a fair-minded retailer, and analyze the influence of three types of fair reference points (supplier's profit reference point, Nash bargaining solution reference point and firm's contribution reference point) on the retailer's decision when the supplier sets wholesale price through wholesale price contract under stochastic demand. The research finds that compared with the retailer's fair-neutral situation, no matter which fair reference point the retailer chooses, it will reduce its order quantity to retaliate against the unfair situation of the supplier. Compared with the selection of the supplier's profit reference point, under the same degree of fair preference, when selecting Nash bargaining reference point and firm's contribution reference point, the retailer will adjust the order quantity reduced by fair preference in reverse according to its own power and contribution. Specifically, when the bargaining power and contribution rate are large, the retailer will reduce the order quantity and increase the degree of retaliation, conversely, the retailer will increase the order quantity and reduce the degree of retaliation. Compared with the firm's contribution reference point, when bargaining power and contribution rate are equal to each other in the same degree of fair preference, the retailer's order quantity under Nash bargaining solution reference point is lower, which means Nash bargaining solution reference point has more influence on its decision than firm's contribution reference point. Finally, the conclusions are numerically analyzed by MATLAB.

Keywords: Fairness reference points · Profits · Power · Contribution

1 Introduction

Behavioral economics believes that, unlike the rational egoists of standard economics, real people are limited rational individuals with morality and emotion. Due to the role of morality and emotion, advocating fairness and reciprocity is the common attribute of most people, and the sense of fairness may play a vital role in bilateral cooperation [1]. In the practice of operation and management, there are many cases of cooperation rupture due to the neglect of the fair demands of the partners. For example, in 2016, Volkswagen shifted its contradictions to two key component suppliers, which made them feel unfair and resisted, leading to the threat of production suspension in many factories including

© The Author(s), under exclusive license to Springer Nature Switzerland AG 2022
H. Han and E. Baker (Eds.): SDSC 2022, CCIS 1725, pp. 49–63, 2022.
https://doi.org/10.1007/978-3-031-23387-6_4

Volkswagen's main plant in Germany[1]. In 2019, Apple supply chain executive Tony Blevins said that Qualcomm had been Apple's sole baseband supplier for the past few years, but its unfair pricing eventually ended the partnership, so Apple chose MediaTek and Samsung as 5G chip suppliers for iPhone 2019[2].To sum up, it can be seen from both theory and practice that fairness has an important impact on the cooperation between enterprises in the supply chain.

Experiments found that members of the supply chain would have fairness concern, focusing not only on the maximization of their own interests, but also on the fairness of profit distribution in the supply chain. If an equity outcome is achieved, the supply chain may achieve a better performance; however, if an inequity outcome is attained, the supply chain always performs worse [2]. Based on the important impact of fairness concern on supply chain enterprise cooperation, since Fehr and Schmidt (2009) introduced fairness concern into the field of supply chain management, more and more scholars have been engaged in studying this topic with fruitful results. In the study, the supply chain members with fairness concern will choose a certain profit as a fairness reference point, and then judge whether they are treated unfairly by comparing the size of their profits and fairness reference point. In the utility function, their fairness concern is often expressed by using the difference between their own profit and fairness reference point. Current research on supply chain management under fairness concern can be roughly divided into three categories according to the selected fairness reference points.

The first is choice the certain multiple of the supply chain members' profits as the fairness reference point. This reference point has the most fruitful research results, and it can be divided into two situations. One is use the multiple of the trading party's profit as a fairness reference point, which is a vertical fairness concern, such as the retailer (supplier) takes a multiple of the supplier's (retailer) profit as the fairness reference point (Cui et al. 2007), or the retailer directly takes the supplier's profit as the reference point (Nie and Du 2017). With the multiple of the transaction party profit as a fairness reference point and linear demand conditions, Cui et al. (2007) found that when the retailer's fairness concern parameters meet certain conditions, the suppliers can also coordinate the supply chain with a simple wholesale price contract. Caliskan-Demirag et al. (2010) extended the demand function to nonlinear cases (2007) based on Cui et al. (2007), and found that it was similar to the conclusion of Cui et al. (2007). Subsequently, Wu and Niederhoff (2014) extended the study of Caliskan-Demirag et al. (2010) to the context of random demand. Another situation is that a certain multiple of third-party profits in the supply chain is used as a fair reference point, which is a horizontal fairness concern. For example, retailers take a certain multiple of the profits of another retailer in the supply chain as a fairness reference point (Ho et al. 2014). In this context, Shu et al. (2018) found that the increase of fairness concern improves remanufacturers' profits when recyclers use the profits of another recycler as a reference point. There is also literature on both vertical and horizontal fairness concern, such as Liu et al. (2018) studied the impact of logistics service providers on order distribution in both directions, and proposed incentive contracts to optimize decisions during order allocation. Pu et al. (2019) studied the impact of vertical and horizontal fairness concern of physical stores on manufacturers' online

---

[1] http://www.xinhuanet.com/world/2016-08/20/c_129243668.htm.
[2] https://www.sohu.com/a/252535725_100110198.

channel models. Liu et al. (2020) studied the impact of service level on the profits of the members of the supply chain in the situations where manufacturers have vertical fairness concern and horizontal fairness concern, and studied the optimal decisions of the members of the supply chain in different situations. In addition, under the first type of fair reference point, many scholars have studied the green supply chain (Huang et al. 2018; Zhou et al. 2020), low-carbon supply chain (Li and Zhao 2015; Zhou et al. 2017; Wu et al. 2020), closed-loop supply chain (Li and Wang 2019; Ma and Hu 2019) and other topics.

The second category is the Nash bargaining solution reference point. Du et al. (2014) believed that the first kind of fair reference point fair preference parameter is an exogenous parameters related to others, which cannot reflect the endogenous strength and contribution of supply chain members and belongs to an absolute fairness, thus affect members of the perception of fair, and put forward in the consideration enterprise respective strength and contribution of Nash bargaining solution as a fairness reference point. Unlike the first type of fairness reference point, the Nash bargaining solutions have equal-symmetry and Pareto-optimal qualities [21, 22], which is a kind of relative fairness. On the basis of considering the fair preference behavior of supply chain members, this fairness reference point can reasonably reflect the internal strength and contribution of members, which is more convincing in practice and easier to be accepted by members. Based on the fairness reference point of Nash bargaining solution, Chen et al. (2017) constructed the utility function under the fairness concern of retailers and established a two-level pricing and order game model. Du and Zhou (2019) established a model for service quality defect guarantee decision in the logistics service supply chain with Nash bargaining fairness concern and analyzed the impact of fairness concern on the optimal strategies, profits and utilities. Jiang et al. (2019) found that under the Nash bargaining solution fairness reference point, the fairness concern can impose an adverse influence on firms' profits and decrease the magnitude of their carbon emission reductions.

The third category is the firm's contribution reference point. Li et al. (2018) argue that the second type of fairness reference point to overcome the defects of absolute fairness and can be more reasonable reflect the comparison of member strength, but in Du et al. (2014), members of the Nash bargaining process is just a fairness concern based on the inner psychological game of supply chain members, depicting the internal fair utility model is not fully mature. Based on this, Li et al. (2018) proposed that the supply chain profit share corresponding to the firm's contribution was taken as a fairness reference point, that is, the firm's contribution reference point, to depict the impact of the strength of supply chain members on fairness, and further enrich and improve the description and characterization of the fairness concern of supply chain members. However, at present, except for Li et al. (2018), few scholars carry out supply chain research based on the third type of fairness reference points.

In conclusion, we found that most studies are based on a certain type of fairness reference points, studying the influence of fair preferences of supply chain members on supply chain decisions and benefits, but the choice of fairness reference points varies, and the corresponding utility function and profit function vary differently [29], and the basis on which members make their decisions will also change. In this regard, people can not help but ask the difference of different fairness reference points on the decision of fair preferences, which kind of fairness reference point is more beneficial for fair preferences, and which kind of fairness reference point is more conducive to the coordination of supply chain. These issues are of great interest for policymakers in the supply chain, but the literature lacks relevant research content. In addition, since the introduction of random distribution function will make the problem complicated, difficult to solve, and difficult to get the analytical solution of the problem. As a result, most of the existing studies are carried out under the condition of deterministic demand, with few research results on fair supply chain management under random market demand. Therefore, under the random demand, when given the wholesale price of a supplier, we study the decision of retailers under different fairness reference points, analyze the impact of the choice of different fairness reference points on retailer's decision, and explore what factors are related to these impact, so as to provide the decision basis for the participants in the supply chain to choose fairness reference points. Finally, the influence of different fairness reference points on the profits of retailers and suppliers is analyzed by numerical examples, and the obtained conclusions are verified.

2 Model Description

This paper considers a supply chain system consisting of a supplier and a retailer, then the supplier produces the product at a certain cost and sells it to consumers through the retailer. Suppliers are the leaders and retailers are the followers, and moreover, for simplification purposes, this paper only considers the behaviors of fairness concern of retailers. The market demand of the product is a random demand, and the sales price is an exogenous variable. Before the demand is realized, the supplier first sets the wholesale price through the wholesale price contract. The retailer either accepts or leaves. If the retailer accepts, it needs to decide the quantity of the order, and after the demand is fulfilled, the retailer sells the product at a fixed price. This paper considers three types of fairness reference points: the supplier's realized profit, Nash bargaining solution of supplier and retailer's Nash bargaining game, and supply chain profit corresponding to retailer contribution as fair reference points, which are called supplier's profit reference point, Nash bargaining solution reference point and firm's contribution reference point respectively. It is assumed that the residual value and penalty cost of stock are not considered during the sales period. The notations involved in this research are given in Table 1.

The model when the supply chain members do not consider the fair preference as the benchmark model (fairness neutral supply chain) is used to analyze the impact of different fairness reference points on the supply chain. First, we consider the situation of the fairness neutral supply chain under the wholesale price contract. The expected profit functions of retailer, supplier and supply chain are:

$$\begin{cases} \pi_r = pS(q) - wq \\ \pi_s = (w - c)q \\ \pi_{sc} = pS(q) - cq \end{cases} \tag{1}$$

Table 1. Variable definition

Notation	Definition
$D > 0$	Random market demand for the sales season
q	Order quantity by the retailer
c	Unit cost of products
w	Wholesale prices of the products
p	Sales price of the products
$S(q)$	Expected sales volume by retailers
λ ($\lambda > 0$)	Retailer's fairness concern coefficient
α ($0 < \alpha < 1$)	The retailer's bargaining power
θ ($0 < \theta < 1$)	The proportion of profit that the retailer thinks its contribution corresponds to the profit of the supply chain
π_r, π_s, π_{sc}	Retailer's profit, supplier's profit, supply chain's profit
u_r, u_s, u	Retailer's utility, supplier's utility, and supply chain's utility
$q_1^*, q_2^*, q_3^*, q_0^*$	Retailer's optimal order quantity under the condition of supplier's profit reference point, Nash bargaining solution reference point, firm's contribution reference point and fairness neutral
$\pi_{r1}^*, \pi_{r2}^*, \pi_{r3}^*$	Retailer's expected profit under the above three fairness reference points
$u_{r1}^*, u_{r2}^*, u_{r3}^*$	Retailer's expected utility under the above three fairness reference points
$F(\cdot)$	Cumulative distribution function of the demand

Referring to the study of Du et al. (2010), the optimal order quantity and wholesale price of the products are, respectively,

$$q_0^* = F^{-1}[1 - \frac{w}{p}], w(q_0^*) = p\overline{F}(q_0^*).$$

Furthermore, the expected profit and utility of the retailer are, respectively,

$$\pi_{r0}^* = u_{r0}^* = pS(q_0^*) - w(q_0^*)q_0^*.$$

Next, we consider the case where the retailer cares about fairness concern.

3 Decisions Under Different Fairness Reference Points

Based on the model developed by Fehr and Schmidt (1999), Cui et al. (2007) first presented the fairness-minded retailer's utility function as following

$$u_r^{FS} = \pi_r - \lambda \max(\gamma \pi_s - \pi_r, 0) - \beta \max(\pi_r - \gamma \pi_s, 0),$$

where π_r is the retailer's monetary payoff, π_s is the supplier's profit, the equitable outcome for the retailer is γ times the supplier's payoff, λ is the retailer's disadvantageous inequality parameter, β is the retailer's advantageous inequality parameters, and $\lambda \geq \beta$, $0 < \beta < 1$.

Note that the supplier has the dominance of the supply chain and allocates the channel profit, while the retailer is a receiver of the supplier's allocation profit. Therefore, it is reasonable to consider that only the retailer is fairness concerned (Li et al. 2018). At same time, some studies have shown that the retailer cares about the disadvantage inequity instead of advantage inequity under supplier Stackelberg. Therefore, we set $\beta = 0$, and for briefly we let $\gamma = 1$, this assumption is also adopted by Du et al. (2010). Further, the utility of the fairness-concerned retailer is

$$u_r^{FS} = \pi_r - \lambda(\pi_s - \pi_r).$$

Following, we will use this utility function to consider the fairness-minded retailer's decisions under different reference point.

3.1 SUpplier's Profit Reference Point

When the retailer chooses to use the supplier's profit as the fairness concern reference point, given the supplier's wholesale price, its utility function is

$$u_{r1} = \pi_r - \lambda(\pi_s - \pi_r) = (1 + \lambda)[pS(q) - wq] - \lambda(w - c)q. \tag{2}$$

Then, the optimal order quantity and wholesale price are

$$q_1^* = F^{-1}[1 - \frac{w}{p} - \frac{\lambda(w - c)}{(1 + \lambda)p}] \quad w(q_1^*) = \frac{p(1 + \lambda)\overline{F}(q_1^*) + \lambda c}{1 + 2\lambda}.$$

Furthermore, the retailer's expected profit and utility are, respectively,

$$\pi_{r1}^* = pS(q_1^*) - w(q_1^*)q_1^*, \quad u_{r1}^* = (1 + \lambda)[pS(q_1^*) - w(q_1^*)q_1^*] - \lambda[w(q_1^*) - c]q_1^*.$$

Property 1. The relationships with disadvantageous inequality parameter λ are as follows:

(1) the optimal order quantity q_1^* decreases with λ;
(2) the relationships between π_{r1}^* and u_{r1}^* and λ are non-monotonic, respectively.

Proof: See the Appendix.

3.2 Nash Bargaining Solution Reference Point

Du et al. (2014) introduced Nash bargaining solution [20–21] into the study of supply chain management for the first time as a fairness reference point, and it is pointed out that using the Nash bargaining solution as the reference point can capture the supply chain members power and contribution, and overcomes the defect of only considering absolute fairness in the past, which is more in line with the reality. Based on this, this paper draws on previous research and assumes that retailer's Nash bargaining solution fairness reference point is $\overline{\pi}_r$.

Given the supplier's wholesale price, the retailer's utility function under Nash bargaining solution reference point is:

$$u_{r2} = \pi_r + \lambda(\pi_r - \overline{\pi}_r) = (1 + \lambda)\pi_r - \lambda\overline{\pi}_r, \tag{3}$$

where $\overline{\pi}_r$ is the optimal solution of the following Nash bargaining problem:

$$\max_{\overline{\pi}_r, \overline{\pi}_s}(u_{r2})^\alpha (u_{s2})^{1-\alpha}$$

$$s.t. \quad \overline{\pi}_r + \overline{\pi}_s = \pi_{sc}$$

$$\overline{\pi}_r, \overline{\pi}_s \in [0, \pi_{sc}]$$

where α and $1 - \alpha$ are the supplier's and retailer's Nash bargaining power, respectively. The solution process is similar to Du et al. (2014), and we can get that the retailer's Nash bargaining solution reference point is $\overline{\pi}_r = \frac{\alpha(1+\lambda)}{1+\alpha\lambda}\pi_{sc}$. Then, the retailer's utility function is

$$u_{r2} = (1 + \lambda)[pS(q) - wq] - \lambda\frac{\alpha(1 + \lambda)}{1 + \alpha\lambda}[pS(q) - cq].$$

Thus, we can get the optimal order quantity and wholesale price are, respectively,

$$q_2^* = F^{-1}[1 - \frac{w}{p} - \frac{\lambda\alpha(w - c)}{p}], \quad w(q_2^*) = \frac{p\overline{F}(q_2^*) + \lambda\alpha c}{1 + \lambda\alpha}.$$

Furthermore, the retailer's expected profit and utility are, respectively,

$$\pi_{r2}^* = pS(q_2^*) - w(q_2^*)q_2^*, \quad u_{r2}^* = (1 + \lambda)[pS(q_2^*) - w(q_2^*)q_2^*] - \lambda[w(q_2^*) - c]q_2^*.$$

Property 2: The relationships with disadvantageous inequality parameter λ and Nash bargaining power α are as follows:

(1) the optimal order quantity q_2^* decreases with λ and α, respectively;
(2) the relationships between $\pi_{r2}^*(u_{r2}^*)$ and λ and α are non-monotonic, respectively.

The proof procedure is similar to Property 1.

3.3 FIrm's Contribution Reference Point

According to Li et al. (2018), we use θ to denote the reference share of retailer's profit in channel profit according to his contribution to the supply chain. This means the retailer regards $\theta\pi_{sc}$ as fair outcome and feels unfair when the percentage of channel profit he occupied is less than $\theta\pi_{sc}$. This reference point reflects how much the member should gain from the total channel payoff.

When the retailer chooses to use his contribution profit as the fairness concern reference point, given the supplier wholesale price, his utility function is

$$u_{r3} = \pi_r - \lambda(\theta\pi_{sc} - \pi_r) = (1 + \lambda)[pS(q) - wq] - \lambda\theta[pS(q) - cq]. \qquad (4)$$

Then, the optimal order quantity and wholesale price of the products are, respectively,

$$q_3^* = F^{-1}[1 - \frac{w}{p} - \frac{\lambda\theta(w - c)}{p(1 + \lambda - \lambda\theta)}], \quad w(q_3^*) = \frac{(1 + \lambda - \lambda\theta)p\overline{F}(q_3^*) + \lambda\theta c}{1 + \lambda}.$$

Furthermore, the retailer's the expected profit and utility are, respectively,

$$\pi_{r3}^* = pS(q_3^*) - w(q_3^*)q_3^*, \quad u_{r3}^* = (1 + \lambda)[pS(q_3^*) - w(q_3^*)q_3^*] - \lambda[w(q_3^*) - c]q_3^*.$$

Property 3: The relationships with disadvantageous inequality parameter λ and the contribution share of retailer's profit in channel profit θ are as follows:

(1) the optimal order quantity q_3^* decreases with λ and θ, respectively;
(2) the relationships between $\pi_{r3}^*(u_{r3}^*)$ and λ and θ are non-monotonic, respectively.

4 Comparing the Optimal Order Quantities

Theorem: When supplier's wholesale price w is given, under the three fairness reference points, supplier's profit reference point, Nash bargaining solution reference point and firm's contribution profit reference point and retailer's fairness neutral condition, the retailer's optimal order quantity satisfies:

(1) when $\frac{\theta}{1+\lambda-\lambda\theta} < \alpha < \frac{1}{1+\lambda}, q_1^* < q_2^* < q_3^* < q_0^*$;
(2) when $\alpha < \frac{\theta}{1+\lambda-\lambda\theta} < \frac{1}{1+\lambda}, q_1^* < q_3^* < q_2^* < q_0^*$;
(3) when $\frac{\theta}{1+\lambda-\lambda\theta} < \frac{1}{1+\lambda} < \alpha, q_2^* < q_1^* < q_3^* < q_0^*$;
(4) when $\frac{1}{1+\lambda} < \frac{\theta}{1+\lambda-\lambda\theta} < \alpha, q_2^* < q_3^* < q_1^* < q_0^*$;
(5) when $\alpha < \frac{1}{1+\lambda} < \frac{\theta}{1+\lambda-\lambda\theta}, q_3^* < q_1^* < q_2^* < q_0^*$;
(6) when $\frac{1}{1+\lambda} < \alpha < \frac{\theta}{1+\lambda-\lambda\theta}, q_3^* < q_2^* < q_1^* < q_0^*$.

Proof: The retailer's optimal order quantities in the four cases are, respectively,

$$q_1^* = F^{-1}[1 - \frac{w}{p} - \frac{1}{1+\lambda}\frac{\lambda(w - c)}{p}], \quad q_2^* = F^{-1}[1 - \frac{w}{p} - \alpha\frac{\lambda(w - c)}{p}],$$

$$q_3^* = F^{-1}[1 - \frac{w}{p} - \frac{\theta}{1+\lambda-\lambda\theta}\frac{\lambda(w - c)}{p}], \quad q_0^* = F^{-1}[1 - \frac{w}{p}].$$

F is monotonically increasing, then F^{-1} is also monotonically increasing, and $w - c > 0$. So, q_1^*, q_2^* and q_3^* are all less than q_0^*. In order to determine the size of q_1^*, q_2^* and q_3^*, we just need to compare the sizes of $1/(1 + \lambda)$, α and $\theta/(1 + \lambda - \lambda\theta)$. Thus, when $\alpha > 1/(1 + \lambda)$, $q_1^* > q_2^*$; otherwise $q_1^* < q_2^*$. When $\theta/(1 + \lambda - \lambda\theta) >$

$1/(1+\lambda)$, which means $\theta > (1+\lambda)/(1+2\lambda)$, $q_1^* > q_3^*$; otherwise $q_1^* < q_3^*$. When $\alpha > \theta/(1+\lambda-\lambda\theta)$, $q_3^* > q_2^*$; otherwise $q_3^* < q_2^*$. So, Theorem (1)–(6) are proved.

According to Theorem 1, given the same wholesale price, the fairness reference points have different effects on retailer's decisions under the different ranges of the parameters λ, α and θ.

Firstly, relative to the fairness neutral situation, no matter retailer chooses which kind of fairness reference point, the optimal order quantity under fairness concern is less than that under fair neutral. And with the increase of λ, α and θ, the gap between fairness-minded quantities and fair-neutral quantity will be bigger and bigger, namely the stronger the retailer's fair preference, bargaining power and contribution rate are, the smaller the order quantity will be in the face of inequity profit distribution from suppliers.

Secondly, compared with the supplier's profit reference point, when the retailer has both fair preference and considers its own power or contribution, under the same degree of fair preference λ, if the bargaining power $\alpha > 1/(1+\lambda)$ and contribution rate $\theta > (1+\lambda)/(1+2\lambda)$, q_2^* and q_3^* are both smaller than q_1^*, and increase with α and θ, respectively; the bigger the parameters, the gap between q_2^* and q_3^* and q_1^* is the larger. Otherwise, if the bargaining power $\alpha < 1/(1+\lambda)$ and contribution rate $\theta < (1+\lambda)/(1+2\lambda)$, q_2^* and q_3^* are both larger than q_1^*, and increase with α and θ, respectively; the bigger the parameters, the gap between q_2^* and q_3^* and q_1^* is the smaller. Further more, when the disadvantageous inequality coefficient λ is the same in the three situations (the strength of disadvantageous inequality aversion is the same), compared with only has the fairness concern situation, the retailer with large bargaining power and contribution rate will further reduce his order quantity; otherwise, the retailer with small bargaining power and contribution rate will increase his order quantity.

Finally, compared the case of the Nash bargaining solution reference point and the firm's contribution reference point. Because of $\theta > \theta/(1+\lambda-\lambda\theta)$, when $\alpha = \theta$, it is obviously that $\alpha > \theta/(1+\lambda-\lambda\theta)$. Further more, we have $q_3^* > q_2^*$. It shows that under the same degree of fair preference, the bargaining power has a greater impact on the retailer's order decision than the firm's contribution rate, that is, the Nash bargaining solution reference point has a greater impact on the retailer's order than the firm's contribution reference point.

To sum up, if the difference of the retailer's orders under fairness concern and fairness neutral is regarded as his punish to the supplier for the inequity profit allocation, then the stronger of the retailer's strength of the fairness concern is, the stronger of the retailer's retaliation is. But compared with only considering the fairness concern case (i.e. the retailer chooses the supplier's profit as fairness reference point), the fairness-minded retailer also considers his bargaining power and contribution to the supply chain at the same time, the retailer's retaliation will be adjusted by his bargaining power and contribution. Specifically, when his bargaining power and contribution rate are large, the retailer will apply a more severe punishment to the supplier, and the penalty increases with the his bargaining power and contribution rate; when his bargaining power and contribution rate are small, the retailer will weaken his punishment due to his weakly bargaining power and small contribution rate, but this degree of weakening will decrease as his bargaining power and contribution rate increase; when retailer's bargaining power and contribution rate parameter are the same, the retailer will generate greater retaliation to the supplier under Nash bargaining solution reference point.

5 Numerical Analysis

In this section, to better illustrate the models proposed previously, we validate the theoretical framework by applying numerical analysis. To make numerical analysis problem tractable and realistic, we take a normal distribution to depict the market demand $D \sim N(1000, 100^2)$, the value of the other parameters of the supply chain are $p = 100$, $w = 60$, $c = 40$. With the help of MATLAB18, We use these examples to illustrate the impact of the parameters on retailer's decision making, profit and utility at different fairness reference points, and on the supplier's profit and supply chain's profit.

5.1 The Impact of Parameter λ

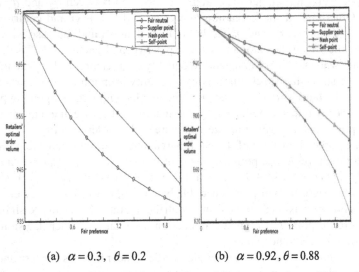

(a) $\alpha = 0.3$, $\theta = 0.2$　　　　(b) $\alpha = 0.92, \theta = 0.88$

Fig. 1. The impact of the disadvantageous inequality coefficient λ on the retailer's order quantity

As can be seen from Fig. 1, given the parameters of bargaining power and contribution rate, the order quantities under the supplier's profit reference point, Nash bargaining solution reference point and firm's contribution reference point are all lower than the fairness-neutral situation, and decreases with λ. When $\alpha = 0.3$ and $\theta = 0.2$, the parameters of bargaining power and contribution rate satisfy $\alpha, \theta / (1 + \lambda - \lambda\theta) < 1 / (1 + \lambda)$, the retailer's bargaining power and contribution rate are small. As a result, the retailer's order quantities under the Nash bargaining solution reference point and the firm's contribution reference point are higher than the order quantity under the supplier's profit reference point, that is $q_1^* < q_2^*, q_3^*$, as shown in Fig. 1(a). When $\alpha = 0.92$ and $\theta = 0.88$, the parameters of bargaining power and contribution rate satisfy $1 / (1 + \lambda) < \alpha, \theta / (1 + \lambda - \lambda\theta)$, the retailer's bargaining power and contribution rate are large. As a result, the retailer's order quantities under the Nash bargaining solution reference point and the firm's contribution reference point are lower than the order quantity under the supplier's profit reference point, that is $q_2^*, q_3^* < q_1^*$, as shown in Fig. 1(b). In addition, when $\alpha = 0.3$ and $\theta = 0.2$, by Fig. 2, the retailer's profits under the three fairness reference points are all lower than his profit under the fairness neutral situation, and the retailer's maximum order quantity under the three situations is $p[1 - F(q_i^*)] = 99.61 > 60 = w$ (where $i = 1, 2, 3$), therefore, his profits decrease with the disadvantageous inequality coefficient λ, which leads to thd supplier's profit and supply chain's profit both decrease. From Fig. 2, although the retailer's profits are declining, his utility is increasing (Fig. 3).

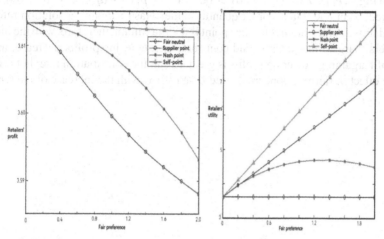

Fig. 2. The impact of the disadvantageous inequality coefficient λ on the retailer's profit and utility

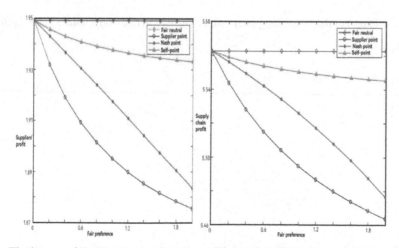

Fig. 3. The impact of the disadvantageous inequality coefficient λ on the supplier's and supply chain's profits

5.2 The Impact of Parameters α and θ

When $\lambda = 0.25$, as is shown in Fig. 4, relative to the supplier's profit reference point, retailer will adjust their order volume according to their bargaining strength α and contribution θ, and the adjustment trend keeps decreasing the order quantity with the increase of α and θ, and when α and θ increase to a certain threshold, the order quantity changes from greater than the supplier's profit reference point to less than; when $\alpha = \theta$, the retailer's order volume under the firm's contribution reference point is always greater than its order volume under the Nash bargaining solution reference point, explain that, relative to the contribution rate, retailers' bargaining power makes their orders fall even more likely, that is, it has a greater impact on its decisions.

From Fig. 4, $p[1 - F(q_2^*)] = 99.61 > 60 = w$ and $p[1 - F(q_3^*)] = 99.61 > 60 = w$ are established, when the retailer order quantity under Nash bargaining solution reference point and firm's contribution reference point is the maximum, therefore, with the increase of retailers' bargaining power α and contribution rate θ, his profits decrease, and the impact of bargaining power on profits is greater than the contribution rate; but different from the effect of fairness concern degree λ on utility, with the increase of the retailer's

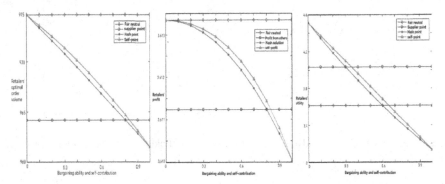

Fig. 4. The impact of the bargaining power α and contribution rate θ on retailer's order, profit and utility

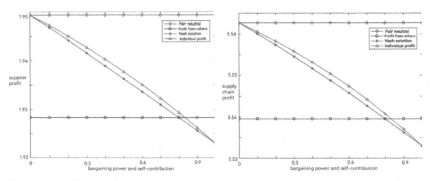

Fig. 5. Impact of bargaining power α and contribution rate θ on supplier's and supply chain's profits

bargaining power α and contribution rate θ, its utility decrease. From Fig. 5, with the increase of the bargaining strength α and the contribution rate θ, the supplier's profits decrease, which eventually leads to the supply chain's profits declining.

6 Conclusion

This paper mainly studies the impact of the retailer's different fairness reference points on the supply chain under the wholesale price contract. We found that, relative to the fair-neutral situation, fairness-minded retailer will retaliate against supplier by reducing orders in the face of inequity profit allocation. If the retailer cares fairness and also has bargaining power or contribute to the supply chain, retailer will adjust the degree of retaliation to suppliers according to his power and contribution rate. Finally, we found that the Nash bargaining solution reference point has more effect on the retailer's decision than the firm's contribution profit reference point under the same the disadvantageous inequality coefficient.

The disadvantage of the paper is, limited by the complexity of the model and the space, this paper only considers the impact of different fairness reference points on the retailer's decision, and the effects of different fairness reference points on supplier's decision should be further studied.

Acknowledgement. This paper was supported by the following fund projects: National Natural Science Foundation of China (No. 71761004); China Humanities and Social Sciences Youth Fund Project of the Ministry of Education (No. 17XJC630006); China Postdoctoral Science Foundation (No. 2017M612868).

Appendix

Proof: (1) Let's $q_1^* = F^{-1}[1 - \frac{w}{p} - \frac{1}{1+\lambda}\frac{\lambda(w-c)}{p}] = F^{-1}[t_1]$. Since the first derivative of function F^{-1} is monotonically increasing, and $w - c$ is greater than zero, judging the sign of the first partial derivative of q_1^* with respect to λ is equivalent to judging the sign of the first partial derivative of t_1 with respect to λ. And $\frac{\partial t_1}{\partial \lambda} = -\frac{w-c}{p(1+\lambda)^2} < 0$, so $\frac{\partial q_1^*}{\partial \lambda} < 0$.

Proof: (2) The first partial derivative of π_{r1}^* with respect to λ can be obtained: $\frac{\partial \pi_{r1}^*}{\partial \lambda} = p\frac{\partial S(q_1^*)}{\partial \lambda} - w\frac{\partial q_1^*}{\partial \lambda} = \frac{\partial q_1^*}{\partial \lambda}\{p[1 - F(q_1^*)] - w\}$. Since $\frac{\partial q_1^*}{\partial \lambda} < 0$, when $p[1 - F(q_1^*)] > w$, $\frac{\partial \pi_{r1}^*}{\partial \lambda} < 0$; otherwise, when $p[1 - F(q_1^*)] < w$, $\frac{\partial \pi_{r1}^*}{\partial \lambda} > 0$. Then the direction of π_{r1}^* is uncertain with respect to λ. The first partial derivative of u_{r1}^* with respect to λ can be obtained:

$$\frac{\partial u_{r1}^*}{\partial \lambda} = [pS(q_1^*) - wq_1^* - (w - c)q_1^*] + \frac{\partial q_1^*}{\partial \lambda}\{(1 + \lambda)\{p[1 - F(q_1^*)] - w\} - \lambda(w - c)\},$$

Since the sign of $\frac{\partial u_{r1}^*}{\partial \lambda}$ with respect to λ is uncertain, the direction of u_{r1} changing with respect to λ is uncertain.

References

Li, W.: Impact of subjective sense of equity on acceptance of redistributive taxation reform: perspective of behavioral economics. Finance Trade Res. **28**(07), 77–87 (2017)

Li, Q., Guan, X., Shi, T., Jiao, W.: Green product design with competition and fairness concerns in the circular economy era. Int. J. Prod. Res. **58**(01), 165–179 (2019)

Fehr, E., Schmidt, K.M.: A theory of fairness, competition, and cooperation. Q. J. Econ. **114**(03), 817–868 (1999)

Cui, T.H., Raju, J.S., Zhang, Z.J.: Fairness and channel coordination. Manag. Sci. **53**(08), 1303–1314 (2007)

Nie, T., Du, S.: Dual-fairness supply chain with quantity discount contracts. Eur. J. Oper. Res. **258**(02), 491–500 (2017)

Caliskan-Demirag, O., Chen, Y.F., Li, J.: Channel coordination under fairness concerns and nonlinear demand. Eur. J. Oper. Res. **207**(03), 1321–1326 (2010)

Wu, X., Niederhoff, J.A.: Fairness in selling to the newsvendor. Prod. Oper. Manag. **23**(11), 2002–2022 (2014)

Ho, T., Su, X., Wu, Y.: Distributional and peer-induced fairness in supply chain contract design. Prod. Oper. Manag. **23**(02), 161–175 (2014)

Shu, Y., Dai, Y., Ma, Z.: Pricing decision in a reverse supply chain with peer-induced fairness concern. Ind. Eng. Manag. **23**(03), 116–122+131 (2018)

Liu, W., Wang, D., Shen, X., Yan, X., Wei, W.: The impacts of distributional and peer-induced fairness concerns on the decision-making of order allocation in logistics service supply chain. Transp. Res. Part E: Logist. Transp. Rev. **116**, 102–122 (2018)

Pu, X., Liu, R., Jin, D.: Manufacturer's distribution strategy considering physical store's fairness concern. Oper. Res. Manag. **28**(11), 178–184 (2019)

Liu, D., Li, D., Zheng, X.: A research on product pricing decision of supply chainconsidering service level under fairness concerns. Nankai Manag. Rev. **23**(01), 98–106+199 (2020)

Huang, H., Yang, D., Yan, Y., Ji, Y.: Closed-loop supply chain pricing decision considering greenness of products under fairness preference. Ind. Eng. Manag. **23**(06), 162–172 (2018)

Zhou, Y., Hu, J., Liu, J.: Decision analysis of a dual channel green supply chain considering the fairness concern. Ind. Eng. Manag. **25**(01), 9–19 (2020)

Li, Y., Zhao, D.: Low-carbonization supply chain coordination with contracts considering fairness preference. J. Manag. Eng. **29**(01), 156–161 (2015)

Zhou, Y., Bao, M., Chen, X., Xu, X.: Co-op advertising and emission reduction cost sharing contract and coordination in low-carbon supply chain based on fairness concerns. China Manag. Sci. **25**(02), 121–129 (2017)

Wu, X., Ai, X., Nie, J.: The impact of asymmetric fairness concern on low carbon supply chain decisions. J. Central Univ. Finance Econ. **02**, 96–105 (2020)

Li, X., Wang, Q.: An optimal decision research on closed-loop supply chain considering retailer's service and fairness concern. Manag. Rev. **31**(04), 230–241 (2019)

Ma, D., Hu, J.: Study on dynamic equilibrium strategy of closed-loop chain with retailers' fair behavior and retailer recycling. China Manag. Sci. **27**(04), 70–78 (2019)

Du, S., Nie, T., Chu, C., Yu, Y.: Newsvendor model for a dyadic supply chain with Nash bargaining fairness concerns. Int. J. Prod. Res. **52**(17), 5070–5085 (2014)

Nash, F.J.: The bargaining problem. Econom. J. Econom. Soc. **18**(02), 155–162 (1950)

Nash, T.: The colorimetric estimation of formaldehyde by means of the Hantzsch reaction. Biochem. J. **55**(03), 416–421 (1953)

Chen, J., Zhou, Y., Zhong, Y.: A pricing/ordering model for a dyadic supply chain with buyback guarantee financing and fairness concerns. Int. J. Prod. Res. **55**(18), 5287–5304 (2017)

Du, N., Zhou, S.: Quality defect guarantee decision in logistics service supply chain with fairness concern. Oper. Res. Manag. **28**(07), 3443 (2019)

Jiang, W., Yuan, L., Wu, L., Guo, S.: Carbon emission reduction and profit distribution mechanism of construction supply chain with fairness concern and cap-and-trade. PLoS ONE **14**(10), e0224153 (2019). https://doi.org/10.1371/journal.pone.0224153

Li, Q., Xiao, T., Qiu, Y.: Price and carbon emission reduction decisions and revenue-sharing contract considering fairness concerns. J. Clean. Prod. **190**, 303–314 (2018)

Du, S., Du, C., Liang, L., Liu, T.: Supply chain contracts and coordination considering fairness concerns. J. Manag. Sci. **13**(11), 41–48 (2010)

Health and Biological Data Science

Health and Biological Time Series

EGRE: Calculating Enrichment Between Genomic Regions

Yang-En Yu and Mary Lauren Benton[✉]

Department of Computer Science, Baylor University, Waco, TX 76798, USA
marylauren_benton@baylor.edu

Abstract. Advances in high-throughput experimental assays have enabled large-scale studies of genome function, especially in the non-protein-coding regions of the genome. These non-coding regions play important roles in the regulation of gene expression, while genetic variation in these regions has been associated with a wide range of complex diseases. In order to annotate and characterize non-coding regions, researchers must integrate data from multiple experimental assays. One common analysis is to quantify the relationship between pairs of genomic regions to determine whether they are enriched for overlap. We present EGRE (Enrichment of Genome REgions), a Python tool to quantify enrichment between pairs of genomic regions using permutation. Our method provides functions to account for common confounders in genomic studies, such as GC-nucleotide content, and calculate different types of overlap between intervals. We demonstrate that our approach returns accurate results using simulated data and recapitulates known relationships from the regulatory genomics literature. Overall, EGRE provides an accurate, flexible, and accessible method to study genomic interval files.

Keywords: Genomics · Functional enrichment · Bioinformatics

1 Introduction

High-throughput experimental assays revolutionized the field of computational genomics, enabling the functional characterization of many parts of the genome. Large consortia, such as ENCODE and Roadmap Epigenomics, have compiled massive databases of functional genomics data across a diverse collection of human cell types and tissues [1–3]. Complementary projects have begun the same process for commonly used model organisms, including worms and fruit flies [4]. Many of the genomics assays used by these consortia produce sets of genomic intervals that require complex data integration techniques in order to infer biological relationships between the annotations. One common analysis compares pairs of genomic annotations. In these analyses, the user is interested in quantifying whether there is overlap, or colocalization, between two sets of regions. Significant overlap could indicate a relationship between the annotated regions, which can inform biologists about the molecular function of the loci in a particular context or generate new biological hypotheses. For example, specific proteins, called transcription factors (TFs), can bind to DNA sequence elements to activate

© The Author(s), under exclusive license to Springer Nature Switzerland AG 2022
H. Han and E. Baker (Eds.): SDSC 2022, CCIS 1725, pp. 67–79, 2022.
https://doi.org/10.1007/978-3-031-23387-6_5

gene expression. YY1 is one such TF that is known to bind gene-regulatory sequences, facilitate chromatin looping, and ultimately influence downstream gene expression [5]. Enrichment between experimentally-defined YY1 binding sites and a set of putative gene-regulatory elements could implicate these regions in the transcriptional activation of a particular cell type or condition.

Earlier methods, such as LOLA [6], calculate enrichment between region sets using a Fisher's exact test. A second class of methods, including GAT, Bedshift, and GLANET, relies on permutation-based approaches to simulate a null distribution and calculate an empirical p-value [7–9]. These methods provide more flexibility to define the appropriate null hypothesis; however, many standalone programs provide limited customization options. For example, GAT can match guanine/cytosine (GC) nucleotide content based on isochores, while GLANET supports isochore matching and exact GC content matching, but neither implement customizable GC ranges [7, 9]. Some programming libraries and packages exist, but are largely inaccessible to users without computational experience. Graphical or web interfaces can bridge this gap [10], but are difficult to incorporate into broader workflows and can be more challenging to reproduce.

Furthermore, in permutation-based enrichment approaches, it is vital to specify an appropriate null model. Simple randomization approaches do not account for known confounders in genomic data, such as uneven distribution across chromosomes or guanine/cytosine (GC) nucleotide content [11]. Existing tools fail to give the user adequate control over the composition of the null model by limiting the types of overlap calculated, or providing limited options to customize the null model. Additionally, current approaches are largely tailored for the analysis of human samples. Users with data from studies of model organisms are unable to use these approaches without substantial editing.

Our approach, EGRE (Enrichment of Genome REgions), overcomes these challenges by providing a simple command-line utility to calculate enrichment between pairs of genomic regions. We support a range of model organisms and genome assemblies, and provide options to control the composition of the null model to ensure the most accurate results, including GC sequence content matching. EGRE is implemented in Python, making it simple to execute on a range of personal or academic systems. We demonstrate that EGRE corrects for common confounders and provides biologically relevant results, without requiring any formal programming knowledge.

2 Results

2.1 EGRE is a Customizable Tool to Calculate Enrichment Between Pairs of Genomic Interval Files

EGRE is available as a command line tool. The goal of EGRE is to determine the significance of overlap between a pair of genome interval files. Input data is expected in the standard BED format [12], with the first three columns representing the chromosome name, start position, and end position of the interval, followed by optional metadata. The statistical significance of overlap between these genomic intervals is quantified by comparing the observed amount of overlap between the regions to an empirical null model generated by a series of random permutations.

When comparing two BED-formatted files, A and B, overlap is calculated by counting the number of base pairs (bp) or elements that are covered by both of the files. Then the intervals in A are held constant, while those from B are randomly shuffled throughout the genome (B'). The shuffled regions maintain the same count and length distribution as the original B regions and are not allowed to overlap after shuffling. From here, EGRE quantifies the overlap between A and B'. Given a large number of random permutations (n > 1000; specified using --iters), EGRE generates an empirical null distribution and compares the observed overlap to the expected amount of overlap, calculating an empirical p-value (Fig. 1).

In addition to this default behavior, we have added a series of features that allow this tool to analyze genome data from a wide range of species, generate customized null distributions, and operate on both individual machines and high-performance computing resources.

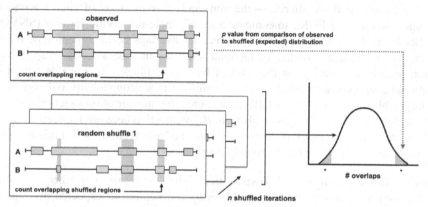

Fig. 1. Schematic showing the workflow of EGRE to compute enrichment between two sets of genomic regions using permutation.

Selecting the Desired Genome Assembly and Species. EGRE's ability to compare sets of genomic regions is widely applicable to a range of biological processes and models. We currently can perform enrichment analyses on human samples and a number of commonly studied model organisms, including mice, fruit flies (*Drosophila melanogaster*) and yeast (*Saccharomyces cerevisiae*). Due to differences in data availability for different reference genomes, we also support the use of multiple human genome assemblies. When running EGRE, the user can choose the desired species and assembly version using the --species argument.

Specifying a Custom Blacklist File. The accuracy of the results is dependent on the ability to construct an accurate null distribution. To facilitate the customization of the null distribution, users can provide a "blacklist" file to specify any genomic locations that should be excluded from the shuffling procedure (--blacklist [file]).

For example, many genomic enrichment analyses use functional genomics data derived from experimental assays such as chromatin immunoprecipitation followed by sequencing (ChIP-seq). In ChIP-seq, certain parts of the genome are prone to experimental bias or error; thus, such regions are filtered from the results and will not be represented in any interval files derived from these data [13]. Providing a list of such blacklisted regions ensures that permutations do generate any random regions overlapping these locations.

By default, EGRE will filter out the blacklisted regions defined by the ENCODE consortium for functional genomics data for the human, mouse, and fly genomes [13].

Measuring Multiple Types of Overlap Between Files. There are multiple ways to quantify overlap between a pair of genomic interval files. By default, EGRE calculates the number of bp shared between the two files. However, in some analyses, the user may want to define overlap differently. We also support element-wise overlap (--elemwise), where overlap between two intervals is considered a hit if they intersect by at least 1 bp.

Due to linkage disequilibrium—the non-random association of alleles across the genome—when using EGRE to compare a set of single-nucleotide variants (SNVs) to another interval file, users may want to consider overlaps between haplotype blocks, rather than individual variants. To facilitate this, we include a --by_hap_block option to quantify overlaps at the level of haplotype blocks. To use this option, an additional column is required in the file containing the genetic variants to designate the haplotype block for each SNV. EGRE then counts the number of intersections between each haplotype block and the intervals from *B*. Intersection between the region and any SNV in the haplotype block is counted as a hit, but intersections with multiple SNVs within the same haplotype block are only counted once.

Printing the Empirical Null Distribution. The --print_counts_to option allows users to print both the observed overlap the number of expected overlaps per iteration to a file. This provides users with the ability to visualize the null distribution and compare both the summary statistics and the full distributions between different parameter settings.

Parallelizing the Permutations. Depending on the size of the file and the complexity of the parameters for the null distribution, it can be time consuming to complete the appropriate number of permutations for each analysis. EGRE provides the ability to multi-thread across a user-specified number of threads using the --num_threads option. This parallelization can be used both on individual machines, or extended to a user-specified number of CPUs on high-performance computing architectures.

2.2 GC Matched Permutations Improve the Accuracy of Enrichment Results

While the default version of EGRE returns biologically relevant results, we aimed to improve the flexibility of the model to allow for increased customization of the null distribution. It is well documented that certain portions of genome, including regulatory elements such as promoters, have higher average GC nucleotide content than other portions of the genome [14]. We observed that random shuffling, even when excluding blacklisted regions, would not be sufficient to account for systemic bias in GC content between sets of genome intervals.

For example, transcription factor (TF) binding is a vital component of the gene regulatory process, where TFs bind to regulatory DNA sequences (such as promoters and enhancers) in order to activate or repress transcription of a particular gene [15]. These TFs bind to particular genomic patterns known as motifs, many of which are GC-rich or located in GC-rich regions [16]. We hypothesized that some sets of genomic intervals, such as TF binding sites, could overlap with other genomic intervals purely because they shared similar levels of GC content, rather than because of a biological association between the regions. Thus, we implemented a GC-matching option for the EGRE framework to ensure that the null distribution could be matched not only for number and length of regions, but also for overall GC content.

Matching Permutations Based on GC Content. The user can specify that the null distribution matches to the GC content of the first interval file, A, by providing the --GC_option flag. If enabled, the option will set the valid range of GC content for permuted regions to be within 10% of the median GC content of A (Fig. 2). To improve flexibility, the upper and lower limits of the GC content range can be set by the user with the --GC_max and --GC_min options.

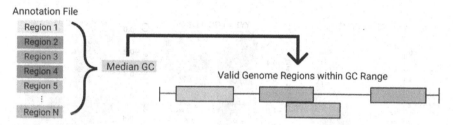

Fig. 2. Schematic showing the filtering step for the GC option before the shuffling step in enrichment calculation. This step filters the tiles created from the whole genome by checking if the tile GC percent is within the valid range of values.

Experiment 1: Comparing Enrichment with GC Matching for TF ChIP-seq and Simulated GC-rich Files. To test the effectiveness of our GC matching we calculated enrichment using EGRE between a series of TF ChIP-seq files (CTCF, YY1, and FOS) and simulated files with similar GC content (Fig. 3).

Since the simulated files were randomly generated, we did not expect a true biological relationship between the TF ChIP-seq and the simulated files. However, because we matched the GC content distribution of the simulated files to that of the TF ChIP-seq, we increased the probability of overlap between our files compared to a completely random set of genomic regions, making these files good candidates for GC matching. Since GC matching restricts the available regions for the null distribution, we hypothesized that the expected amount of overlap would increase, reducing the fold change in the result.

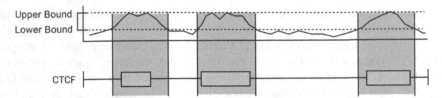

Fig. 3. Schematic showing the generation of simulated GC files for Experiment 1. Regions were chosen from tiles in the genome that had GC content matching the middle 50% of the real file. For example, the CTCF ChIP-seq (median 49% GC) in the right atrium was matched with random regions from tiles with 41–59% GC content.

For each of the tested TF ChIP-seq files, the amount of overlap expected with the permuted intervals increased after GC matching. This resulted in a decreased fold change between the default and matched runs. For CTCF ChIP-seq in the right atrium, we initially observed no significant enrichment, consistent with a lack of biological association between the files; however, after GC matching, we observed a slight depletion of overlap compared to a matched background (Fig. 4A). For YY1 and FOS, the depletion for overlap was reduced, and the fold change moved closer to 0, after we applied the GC correction (Fig. 4B-C).

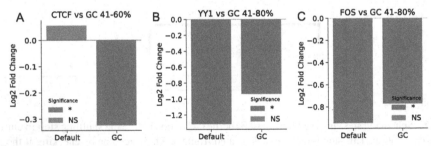

Fig. 4. GC correction applied to experiments with artificial data. The log2-transformed fold change (y-axis) of the comparison between (A) CTCF ChIP-seq (*Homo sapiens* right atrium auricular region tissue), (B) YY1 ChIP-seq (*Homo sapiens* K562 cells), (C) FOS ChIP-seq (*Homo sapiens* K562 cells) and artificially generated genome regions that are similar to the TF ChIP-seq files in GC content. The '*' indicates the results is below the 0.05 p-value threshold while 'NS' indicates the result is above 0.05 p-value threshold and thus considered insignificant result.

Experiment 2: Comparing Enrichment with GC Matching Between TF ChIP-seq and Experimentally-Derived Gene Regulatory Regions. After testing EGRE with simulated data, we next wanted to apply the tool to real sets of genomic regions, where we had clear expectations about the results based on prior literature. We chose TF binding sites from ChIP-seq (for CTCF and YY1), and two other files that are associated with regulatory activity: ChIP-seq for the histone modification H3K27ac, and the set of candidate *cis*-regulatory elements (cCREs) defined by the ENCODE consortium. We also tested two different biological contexts: a primary tissue sample from the right atrium auricular region and the K562 cell line.

We performed three separate comparisons in this experiment: (1) CTCF ChIP-seq vs. cCREs in K562 cells, (2) YY1 ChIP-seq vs. cCREs in K562 cells, and (3) CTCF vs. H3K27ac ChIP-seq in human right atrium. In all three cases, we expected enrichment for TF binding within the putative gene regulatory regions [5]. Due to the composition of the TF binding motifs and the bias towards higher GC-content in gene regulatory regions overall, we also expected the GC matching to reduce the level of fold change by increasing the expected amount of overlap in the null distribution. As expected, we found a 6.4-fold enrichment for CTCF binding sites in cCREs in K562 cells (Fig. 5). We retain the significant enrichment after adjusting for GC content, although the fold change goes down (fold change = 5.6).

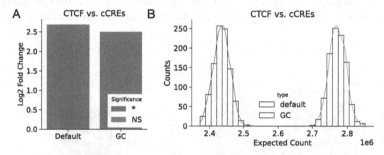

Fig. 5. CTCF binding is enriched in cCREs. (A) The log2 fold change for CTCF binding in H3K27ac regions with and without the GC content correction.(B) The observed overlap is 15,657,762 bp, which is higher than the expected overlap. We observe a lower expected overlap in both cases. The expected overlap distribution shifts when the GC option is enabled in the enrichment calculation. We observe an increase in the expected overlap. The '*' indicates the results is below the 0.05 p-value threshold while 'NS' indicates the result is above 0.05 p-value threshold and thus considered insignificant result.

Similarly, we observe an enrichment of 6.6-fold for YY1 binding in K562 cCRES with the default settings and a 3.6-fold enrichment with the GC option (Fig. 6). The fold change drops more than the CTCF analysis after adjusting for GC content, possibly because of the higher GC content of the YY1 binding sites compared to the CTCF sites (YY1 median 64% v. CTCF median 49%).

Finally, we also observed a significant enrichment for CTCF binding in regions with H3K27ac signal (Fig. 7). This enrichment increases slightly after adjusting for GC content because the amount of expected overlap decreases.

2.3 Parallelization Using Multithreading Reduces the Elapsed Time for Analysis

Due to the number of iterations required to achieve reliable results from EGRE, we wanted to understand the impact of customizations to the null model, such as GC matching, on the amount of time it takes a user to complete the analysis. Users may need to run

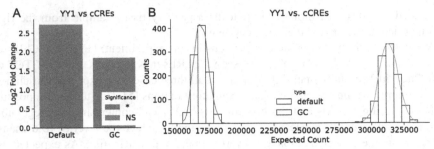

Fig. 6. YY1 binding is enriched in cCREs. (A) The log2 fold change for YY1 binding in cCREs with and without the GC content correction. (B) The observed overlap is 1,115,490, which is higher than the expected overlap. We observe a lower expected overlap in both cases. The expected overlap distribution shifts when the GC option is enabled in the enrichment calculation. We observe an increase in the expected overlap. The '*' indicates the results is below the 0.05 p-value threshold while 'NS' indicates the result is above 0.05 p-value threshold and thus considered insignificant result.

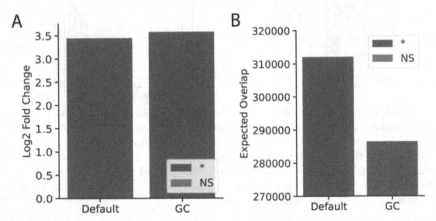

Fig. 7. CTCF is enriched in regions with H3K27ac signal. (A) The log2 fold change for CTCF binding in H3K27ac regions with and without the GC content correction. (B) The expected overlap changes in when the GC option is enabled in the enrichment calculation. We observe a decrease in expected overlap, suggesting that with the addition of GC content constraints in the enrichment calculation. The '*' indicates the results is below the 0.05 p-value threshold while 'NS' indicates the result is above 0.05 p-value threshold and thus considered insignificant result.

the same analysis for multiple factors and different parameter settings; thus, the required analysis time is an important consideration.

To benchmark the performance of the new tool with the GC option enabled, we performed runs with varying numbers of iterations and threads to assess the elapsed running time. We consider the elapsed time in terms of wall-clock seconds. From our benchmarking analysis, we observed that the runtime decreases as number of threads increases (Fig. 8A), and increases in polynomial time as iterations increases (Fig. 8B). It takes less than 30 min to run at least 1000 iterations, regardless of the number of threads used. On average, it takes 10 min to run 1,000 iterations with 20 threads and 76 min to run 10,000 iterations with 20 threads. On an individual machine, we observed only marginally improvement after increasing to more than 20 threads.

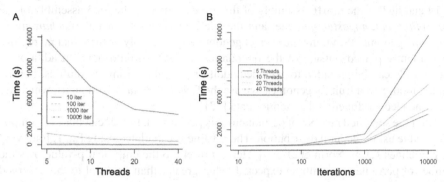

Fig. 8. Performance of the new tool with the GC option enabled across different combinations of threads and iterations. (A) Elapsed time decreases as the number of threads increases. Colored lines designate different numbers of iterations. (B) Elapsed time increases with the number of iterations. Colored lines designate different numbers of threads.

3 Methods

3.1 Implementation of EGRE

EGRE is implemented in Python (version 3) with a command line interface. To manipulate BED-formatted files, we use pybedtools [17], a python wrapper for the command line utility BEDTools [18].

EGRE expected input in BED format, where the first three columns are the chromosome name, start position, and end position of the interval. Additional data can be provided in the BED files, but will not be considered. The only exception is when using the --by_hap_block option to quantify overlap by haplotype block. In this case, the last column of the file should be a number to designate which haplotype block each region belongs to. Similar to the element-wise option, where any overlap between regions is counted as a single "hit", when the --by_hap_block option is enabled, EGRE uses the haplotype ID column to count any overlap regions in the haplotype block as a single hit.

Briefly, the tool calculates overlap between genomic regions by either counting the number of base pairs (bp) or elements that are covered by intervals in both of the files. To generate the null distribution of overlap, the intervals in the first file, A, are held constant, while those from B are randomly shuffled throughout the genome (B'). The number and length of the intervals in B are maintained in B', and are not allowed to overlap after shuffling. If two shuffled regions overlap during a random permutation, one of the regions is randomly shuffled to a new location. The process of reshuffling to avoid overlapping regions in the null distribution can be performed up to 1000 times for each random region. Additionally, the distribution of intervals across chromosomes is maintained by only permuting regions within a single chromosome.

The universe of possible regions for the shuffle regions is dictated by the species and reference genome assembly. We currently support two versions of the human genome (hg19 and hg38), the mm10 assembly of the mouse genome, the dm3 assembly of the *Drosophila melanogaster* genome, and the sacCer3 assembly of the *Saccharomyces cerevisiae* genome. Providing the correct genome and assembly ensures that no regions are randomly placed outside of the boundaries of each chromosome. In addition, a blacklist file can be provided to excluded additional regions from the shuffles. For the human, mouse, and fruit fly genomes, we use the ENCODE blacklist regions, which are recommended for functional genomics data [13].

After the specified number of permutations is performed, EGRE calculates an empirical p-value using that null distribution. The p-value is calculated using Eq. 1, following the guidelines set by North et al. [19]. Here, p refers to the empirical p-value, r is the number of permutations with an expected value greater than or equal to the observed value, and n is the total number of permutations.

$$p = (r + 1)/(n + 1) \tag{1}$$

EGRE reports the amount of observed overlap, the average amount of expected overlap, the standard deviation of the expected distribution, the ratio of the observed to expected overlap (fold change), and the empirical p-value. The full null distribution can be printed to a separate file using the `--print_counts_to` option.

3.2 Implementation of GC Matching for Shuffled Genomic Regions

The GC content option is implemented using the pybedtools wrapper for BEDTools [17, 18]. The GC content option, when selected, creates fixed-size (100 bp), tiles across the entire genome (overlapping at intervals of 50 bp), and calculates the GC content of each tile (Fig. 9).

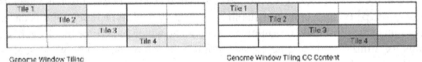

Fig. 9. Schematic showing the tile generation used in the GC option for enrichment calculation. The tiles are 100 bp per window with 50 bp overlap, but the window size and overlap size can be specified by the user to adjust to customized runs. The generated windows are then coupled with genome build's full genome FASTA file as parameters for the `nucleotide_content` function in pybedtools to calculate the GC content of each tile. Darker shades of green indicate higher GC content.

We also calculate the GC content of each entry in the annotation BED file provided to the script. The GC content of each entry in the annotation file is calculated using the `nucleotide_function` in pybedtools, and the median GC percent is found using the median function in NumPy. By default, the GC content range is set as 10% above and 10% below the median of the input set for filtering the previously calculated genome tiles. The genome tiles with GC content within the chosen range are selected as valid regions for shuffling to create the null distribution.

The range used to filter genome tiles can also be changed to custom values by the user to fit customized runs. The filtered regions are then combined with genome blacklist regions as input into the shuffling step in the enrichment calculation.

3.3 Functional Genomics and Simulated Data Used for Experiments

We downloaded ChIP-seq peaks from ENCODE for CTCF, FOS, and YY1 transcription factors, the H3K27ac histone modification (ENCFF233DXO, ENCSR668EVA, ENCSR000EWF, ENCSR946WBN, ENCSR000EGM), and cCREs defined by the SCREEN Registry of cCREs (v3) [20]. These were stored in a BED format, where the first three columns represent the chromosome, start position, and end position for the genomic interval. We confirmed that all ENCODE blacklisted regions were filtered out of the files prior to analysis.

For the artificial datasets, we generated files of equal size, by randomly sampling intervals from the genome with a similar GC content. To determine the range of GC content for sampling, we matched the GC content distribution of the middle 50% of the real data.

3.4 Benchmark Analysis

We used a Dell R740 server with 64 GB of RAM and two Xeon 12 Core, hyperthreaded processors (2.2Ghz) for our benchmarking analysis. We ran no other processes during the benchmarking. We tested EGRE with 16 different combinations of permutations and threads and quantified the elapsed real time in seconds.

3.5 Code Availability

EGRE is licensed under BSD-3, and is available on GitHub: https://github.com/ben tonml/genomic_enrichment.

4 Conclusions

In this work, we present EGRE as a tool to quantify enrichment between pairs of genomic interval files. EGRE has a simple command-line interface that allows for flexibility to generate an accurate null model using permutations, but does not require any programming knowledge to use. In a series of experiments, we show that EGRE can account for common confounders in genomic data, including matching for GC content. In experiments with simulated data, EGRE is able to generate the expected results, and while experiments with real experimental datasets recapitulate known relationships from prior work. Our implementation supports multiple overlap styles and genome assemblies, increasing utility for both human and model organism studies.

EGRE is currently limited to a subset of commonly used species and a single resolution for GC content matching (100 bp). In the future, we plan to add support for additional species and genome assemblies, as well as precomputed GC and blacklist filess to speed up compute time and provide additional customization options for users. Further optimization could include writing this tool in C++ or Cython to improve speed. Ultimately, we believe EGRE is a valuable addition to any bioinformatics toolkit.

References

1. ENCODE Project Consortium, An integrated encyclopedia of DNA elements in the human genome. Nature **489**(7414), 57–74 (2012).https://doi.org/10.1038/nature11247
2. Kundaje, A., et al.: Integrative analysis of 111 reference human epigenomes. Nature **518**(7539), 7539 (2015). https://doi.org/10.1038/nature14248
3. Davis, C.A., et al.: The encyclopedia of DNA elements (ENCODE): data portal update. Nucleic Acids Res. **46**(D1), D794–D801 (2018). https://doi.org/10.1093/nar/gkx1081
4. Celniker, S.E., et al.: Unlocking the secrets of the genome. Nature **459**(7249), 927–930 (2009). https://doi.org/10.1038/459927a
5. Weintraub, A.S., et al.: YY1 is a structural regulator of enhancer-promoter loops. Cell **171**(7), 1573–1588.e28 (2017). https://doi.org/10.1016/j.cell.2017.11.008
6. Sheffield, N.C., Bock, C.: LOLA: enrichment analysis for genomic region sets and regulatory elements in R and bioconductor. Bioinformatics **32**(4), 587–589 (2016). https://doi.org/10.1093/bioinformatics/btv612
7. Heger, A., Webber, C., Goodson, M., Ponting, C.P., Lunter, G.: GAT: a simulation framework for testing the association of genomic intervals. Bioinformatics **29**(16), 2046–2048 (2013). https://doi.org/10.1093/bioinformat-ics/btt343
8. Gu, A., Cho, H.J., Sheffield, N.C.: Bedshift: perturbation of genomic interval sets. Genome Biol **22**, 238 (2021). https://doi.org/10.1186/s13059-021-02440-w
9. Otlu, B., Firtina, C., Keleş, S., Tastan, O.: GLANET: genomic loci annotation and enrichment tool. Bioinformatics **33**(18), 2818–2828 (2017). https://doi.org/10.1093/bioinformatics/btx326
10. Simovski, B., et al.: Coloc-stats: a unified web interface to perform colocalization analysis of genomic features. Nucleic Acids Res. **46**(W1), W186–W193 (2018). https://doi.org/10.1093/nar/gky474
11. Teng, M., Irizarry, R.A.: Accounting for GC-content bias reduces systematic errors and batch effects in ChIP-seq data. Genome Res **27**(11), 1930–1938 (2017). https://doi.org/10.1101/gr.220673.117
12. BED format. https://genome.ucsc.edu/FAQ/FAQformat.html#format1
13. Amemiya, H.M., Kundaje, A., Boyle, A.P.: The ENCODE blacklist: identification of problematic regions of the genome. Sci. Rep. **9**(1), 9345 (2019). https://doi.org/10.1038/s41598-019-45839-z
14. Lenhard, B., Sandelin, A., Carninci, P.: Metazoan promoters: emerging characteristics and insights into transcriptional regulation. Nat. Rev. Genet. **13**(4), 4 (2012). https://doi.org/10.1038/nrg3163
15. Shlyueva, D., Stampfel, G., Stark, A.: Transcriptional enhancers: from properties to genome-wide predictions. Nat. Rev. Genet. **15**(4), 272–286 (2014). https://doi.org/10.1038/nrg3682
16. Wang, J., et al.: Sequence features and chromatin structure around the genomic regions bound by 119 human transcription factors. Genome Res. **22**(9), 1798–1812 (2012). https://doi.org/10.1101/gr.139105.112

17. Dale, R.K., Pedersen, B.S., Quinlan, A.R.: Pybedtools: a flexible python library for manipulating genomic datasets and annotations. Bioinformatics **27**(24), 3423–3424 (2011). https://doi.org/10.1093/bioinformatics/btr539

18. Quinlan, A.R., Hall, I.M.: BEDTools: a flexible suite of utilities for comparing genomic features. Bioinformatics **26**(6) 841–842 (2010). https://doi.org/10.1093/bioinformatics/btq033

19. North, B.V., Curtis, D., Sham, P.C.: A note on the calculation of empirical P values from Monte Carlo procedures. Am. J. Hum. Genet. **71**(2), 439–441 (2002)

20. The ENCODE Project Consortium, Moore, J.E., Purcaro, M.J. et al. Expanded encyclopaedias of DNA elements in the human and mouse genomes. Nature **583**, 699–710 (2020). https://doi.org/10.1038/s41586-020-2493-4

THSLRR: A Low-Rank Subspace Clustering Method Based on Tired Random Walk Similarity and Hypergraph Regularization Constraints

Tian-Jing Qiao, Na-Na Zhang, Jin-Xing Liu, Jun-Liang Shang, Cui-Na Jiao, and Juan Wang(✉)

School of Computer Science, Qufu Normal University, Rizhao 276826, China
wangjuansdu@163.com

Abstract. Single-cell RNA sequencing (scRNA-seq) technology furnishes us with a certainly forceful tool for exploring biological mechanisms from the perspective of single-cell. By clustering scRNA-seq data, different types of cells can be effectively distinguished, which is helpful for disease treatment and the discovery of new cell types. Nevertheless, the existing clustering methods still cannot achieve satisfactory results attributed to the complexity of high-dimensional noisy scRNA-seq data. Therefore, we propose a clustering method called Hypergraph regularization sparse low-rank representation with similarity constraint based on tired random walk (THSLRR). Specifically, the sparse low-rank model rebuilds spatial information from a suite of high-dimensional subspaces by mapping data into subspaces, and removes superfluous information and errors in scRNA-seq data. The hypergraph regularization explores the higher-order manifold structure embedded in the scRNA-seq data. Meanwhile, the similarity constraint based on tired random walk can farther upgrade the learning ability and interpretability of the model. Then, the learned similarity matrix could be for spectral clustering, visualization and identification of marker genes. Compared with other advanced methods, the clustering results of the THSLRR method are more robust and accurate.

Keywords: scRNA-seq · Single-cell type identification · Hypergraph regularization · Similarity constraint

1 Introduction

In the past few years, advances in single-cell RNA sequencing (scRNA-seq) technology have provided a new window of opportunity to learn about biological mechanisms at the single-cell level, and guide scientists in exploring gene expression profiles at the single-cell level [1, 2]. By mining and analyzing scRNA-seq data, we can research cell heterogeneity and identify subgroups. The identification of cell types from scRNA-seq data facilitates the extraction of meaningful biological information, as a matter of unsupervised clustering. With the clustering model, cells that are highly similar will be

H. Han and E. Baker (Eds.): SDSC 2022, CCIS 1725, pp. 80–93, 2022.
https://doi.org/10.1007/978-3-031-23387-6_6

grouped into the same cluster. Because of biological factors and technical limitations, however, scRNA-seq data tend to be high-dimensional, sparse and noisy. Consequently, classical clustering methods like K-means [3] and Spectral Clustering (SC) [4] are no longer suitable for scRNA-seq data, and reliable clustering cannot always be used for downstream analysis.

At present, in order to iron out the difficulties existing in scRNA-seq data clustering research, scholars have put forward numerous clustering methods. For instance, through the in-depth research of shared nearest neighbors, Xu and Su came up with a quasi-cluster-based clustering method (SNN-Cliq), which shows greater superiority in clustering high-dimensional single-cell data [5]. Based on the profound study of multi-kernal learning, Wang et al. proposed the SIMLR method, working out dimensionality reduction as well as clustering of data [6]. Park et al. proposed the MPSSC method, in which the SC framework is modified by adding sparse structure constraint, and the similarity matrix is constructed by using multiple double random affinity matrices [7]. Jiang et al. took into account paired cell differentiability correlation and variance, then proposed the Corr model [8].

At the same time, researchers have also proposed a number of subspace clustering methods and proved that the similarity obtained by the subspace clustering method based on low-rank representation (LRR) is more robust than the pairwise similarity involved in the methods mentioned above [9, 10]. For example, Liu et al. proposed the LatLRR method, integrating feature extraction and subspace learning into a unified framework to better cope with severely corrupted observation data [11]. Zheng et al. presented the SinNLRR method, a low-rank based clustering method, that fully exploits the global information of the data by imposing low-rank and non-negative constraints on the similarity matrix [10]. In order to explore the local information of the data, Zhang et al. proposed the SCCLRR method based on SinNLRR with the addition of local feature descriptions to capture both global and local information of the data [9]. Zheng et al. proposed the AdaptiveSSC method based on subspace learning to figure out the matters of noise and high dimensionality in single-cell data, achieving improved performance on multiple experimental data sets [12].

In this paper, we propose a single-cell clustering method called Hypergraph regularization sparse low-rank representation with similarity constraint based on tired random walk (THSLRR), which aims to capture the global structure and local information of scRNA-seq data simultaneously in subspace learning. Concretely, on the basis of the sparse LRR model, the hypergraph regularization based on manifold learning is introduced to mine the complex high-order relationship in scRNA-seq data. At the same time, the similarity constraint based on tired random walk (TRW) further improves the learning ability of model. The final sparse low-rank symmetric matrix Z^* obtained by THSLRR is further operated to learn the affinity matrix H, then H is used for single-cell spectral clustering, t-distributed stochastic neighbor embedding (t-SNE) [13] visual analysis of cells and genes prioritization. Figure 1 illustrates the specific process and applications of THSLRR.

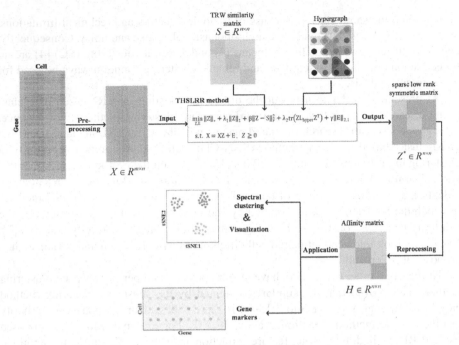

Fig. 1. The framework of THSLRR for scRNA-seq data analysis.

2 Method

2.1 Sparse Low-Rank Representation

The LRR model is a progressive subspace clustering method, which is widely used in data mining, machine learning and other fields. Finding the lowest rank representation of data on the basis of the given data dictionary is the central objective of LRR [14]. Given the scRNA-seq data matrix $X = [X_1, X_2, \ldots, X_n] \in R^{m \times n}$, where m represents the number of genes and n is the number of cells, its LRR formula is expressed as follows:

$$\min_{Z,E} \|Z\|_* + \gamma \|E\|_{2,1} \ s.t. \ X = XZ + E. \tag{1}$$

There, $\| * \|_*$ represents the kernel norm of the matrix, $\| * \|_{2,1}$ is the $l_{2,1}$ norm. E is the error item and Z is the coefficient matrix that demands to be optimized to achieve the lowest rank. $\gamma > 0$ is the parameter to coordinate the influence of errors.

The sparse representation model obtains the sparse coefficient matrix that unravels the close relationship between the data points, what is equivalent to solving the following optimization problem:

$$\min_{Z} \|Z\|_1 \ s.t. \ X = XZ, \tag{2}$$

where $\| * \|_1$ is the l_1 norm. We further combine sparse and low-rank constraints for the extraction of salient features and noise removal to obtain the sparse LRR of the matrix, as follows:

$$\min_{Z,E} \|Z\|_* + \lambda \|Z\|_1 + \gamma \|E\|_{2,1} \ s.t. \ X = XZ + E. \tag{3}$$

Here, λ and γ are regularization parameters.

2.2 Hypergraph Regularization

Extracting local information from high-dimensional sparse noisy data is also a problem worth considering. Therefore, we exploit the hypergraph to encode higher-order geometric relationships among multiple sample points, which can more fully extract the underlying local information of scRNA-seq data.

For a given hypergraph $G = (V, E, W)$, $V = \{v_1, v_2, \ldots, v_n\}$ is the collection of vertexes, $E = \{e_1, e_2, \ldots, e_r\}$ is the collection of hyperedges, W is the hyperedge weight matrix. The incidence matrix R of the hypergraph G is calculated as follows:

$$R(v, e) = \begin{cases} 1 \ if \ v \in e \\ 0 \ others \end{cases} \tag{4}$$

The weight $w(e_i)$ of hyperedge e_i is obtained by the following formula:

$$w(e_i) = \sum_{\{v_i, v_j\} \in e_i} exp^{-\frac{\|v_i - v_j\|_2^2}{\delta^2}}, \tag{5}$$

where $\delta = \sum_{\{v_i, v_j\} \in e_i} \|v_i - v_j\|_2^2 / k$, and k represents the number of nearest neighbors of each vertex. The degree $d(v)$ of vertex v is as follows:

$$d(v) = \sum_{e \in E} w(e)R(v, e). \tag{6}$$

The degree $g(e)$ of hyperedge e is as follows:

$$g(e) = \sum_{v \in V} R(v, e). \tag{7}$$

Then, we obtain the non-normalized hypergraph Laplacian matrix L_{hyper}, as shown below:

$$L_{hyper} = D_v - RW_H(D_H)^{-1}R^T. \tag{8}$$

where vertex degree matrix D_v, hyperedge degree matrix D_H and hyperedge weight matrix W_H are diagonal matrices, and the elements on the diagonal are $d(v)$, $g(e)$ and $w(e)$ respectively.

Under certain conditions of the mapping, z_i and z_j are the mapping representations of the original data points x_i and x_j under the new basis, then the target formula of the hypergraph regularization constraint is as follows:

$$\min_Z \frac{1}{2} \sum_{e \in E} \sum_{(i,j) \in e} \frac{w(e)}{g(e)} \|z_i - z_j\|^2 = \min_Z tr\left(Z\left(D_v - RW_H(D_H)^{-1}R^T\right)Z^T\right)$$

$$= \min_Z tr\left(ZL_{hyper}Z^T\right) \tag{9}$$

2.3 Tired Random Walk

The TRW model was proposed in [15] and proved to be a practical measurement of nonlinear manifold [16]. Therefore, the similarity constraint can not only improve the learning ability of the model for the overall geometric information of the data, but also ensure the symmetry of the similarity matrix, so that the model has better interpretability.

For an undirected weight graph with n vertexes, the transition probability matrix of the random walk is $P = D^{-1}W$, W represents the affinity matrix of the graph, D represents the diagonal matrix with $D_{ii} = \sum_{j=1}^{n} W_{ij}$. According to [17], the cumulative transition probability matrix is $P_{TRW} = \sum_{s=0}^{\infty}(\tau P)^s$ for all vertices, where $\tau \in (0, 1)$ and the eigenvalue of P is at $[0, 1]$, so the TRW matrix is as follows:

$$P_{TRW} = \sum_{s=0}^{\infty}(\tau P)^s = (1 - \tau P)^{-1}. \tag{10}$$

In order to weaken the effect of errors existing in the primary samples and ensure that the paired sample points have consistent correlation weights, we further symmetrize P_{TRW} to obtain final TRW similarity matrix $S \in R^{n \times n}$ as follows:

$$S(x_i, x_j) = \frac{(P_{TRW})_{ij} + (P_{TRW})_{ji}}{2}. \tag{11}$$

2.4 Objective Function of THSLRR

THSLRR learns the expression matrix $Z \in R^{n \times n}$ from the scRNA-seq data matrix $X = [X_1, X_2, \ldots, X_n] \in R^{m \times n}$ with m genes and n cells by the following objective function (12):

$$\min_{Z,E} \|Z\|_* + \lambda_1\|Z\|_1 + \lambda_2 tr\left(ZL_{hyper}Z^T\right) + \beta\|Z - S\|_F^2 + \gamma\|E\|_{2,1}$$

$$s.t. \ X = XZ + E, Z \geq 0, \tag{12}$$

where Z is the coefficient matrix to be optimized, $L_{hyper} \in R^{n \times n}$ is the hypergraph Laplacian matrix, $S \in R^{n \times n}$ is the symmetric cell similarity matrix generated by TRW, $E \in R^{m \times n}$ represents the errors term, $\| * \|_F$ is the Frobenius norm of the matrix, $\lambda_1, \lambda_2,$ β and γ are the penalty parameters.

2.5 Optimization Process and Spectral Clustering of THSLRR Method

The objective function that has multiple constraints of THSLRR is a convex optimization problem. In order to effectively work out the problem (12), we adopt the Linearized Adaptive Direction Method with Adaptive Penalty (LADMAP) [18].

Initially, to separate the objective function (12) by using an auxiliary variable J, and then obtain formula (13):

$$\min_{Z,E,J} \|Z\|_* + \lambda_1\|J\|_1 + \lambda_2 tr\left(ZL_{hyper}Z^T\right) + \beta\|Z - S\|_F^2 + \gamma\|E\|_{2,1}$$
$$s.t.\ X = XZ + E, Z = J, Z \geq 0. \tag{13}$$

Then, the augmented lagrangian multiplier method is introduced to eliminate the linear constraints existing in (13). Therefore, we get the following formula:

$$L(Z, E, J, Y_1, Y_2) = \|Z\|_* + \lambda_1\|J\|_1 + \lambda_2 tr\left(ZL_{hyper}Z^T\right) + \beta\|Z - S\|_F^2 + \gamma\|E\|_{2,1}$$
$$+ \langle Y_1, X - XZ - E\rangle + \langle Y_2, Z - J\rangle$$
$$+ \frac{\mu}{2}\left(\|X - XZ - E\|_F^2 + \|Z - J\|_F^2\right). \tag{14}$$

Here, μ is a penalty parameter, Y_1 and Y_2 are lagrangian multipliers.

Finally, the optimization problem is ironed out by updating one of the variables by turn while fixing the other variables. Therefore, the update rules of Z, E, and J are as follows:

$$Z_{k+1} = \theta_{\frac{1}{\eta\mu}}\left(Z_k - \frac{\nabla_Z q(Z_k)}{\eta}\right). \tag{15}$$

$$E_{k+1}(i, :) = \begin{cases} \frac{\|p_i\| - \frac{\gamma}{\mu_k}}{\|p_i\|}p_i & , \frac{\gamma}{\mu_k} < \|p_i\|. \\ 0, & otherwise \end{cases} \tag{16}$$

$$J_{k+1} = max\left\{\theta_{\frac{\lambda}{\mu_k}}\left(Z_{k+1} + Y_2^k/\mu_k\right), 0\right\}. \tag{17}$$

The sparse low-rank symmetric matrix Z^* is obtained with our THSLRR method, and the elements on both sides of the main diagonal of the matrix Z^* correspond to the similarity weights of the data sample points. Inspired by [19], we use the main direction angle information of matrix Z^* to learn the affinity matrix H. Finally, we use learned matrix H as the input of SC method to obtain the clustering results.

3 Results and Discussion

3.1 Evaluation Measurements

In the experiment, two commonly used indicators are used to assess the effectiveness of THSLRR, namely adjusted rand index (ARI) [20] and normalized mutual information (NMI) [21]. The value of ARI belongs to $[-1, 1]$ while the value of NMI is $[0, 1]$.

Given the real cluster label $T = \{T_1, T_2, \ldots, T_K\}$ and the predicted cluster label $Y = \{Y_1, Y_2, \ldots, Y_K\}$ of n sample points. The formula of ARI is as follows:

$$ARI(T, Y) = \frac{\binom{n}{2}(a_{ty} + a) - [(a_{ty} + a_t)(a_{ty} + a_y) + (a_t + a)(a_y + a)]}{\binom{n}{2} - [(a_{ty} + a_t)(a_{ty} + a_y) + (a_t + a)(a_y + a)]}. \tag{18}$$

Here, a_{ty} denotes the number of data points put in the same class, whereas a_t denotes the number of data points in the same class T but separate Y classes. a_y represents the number of data point pairs that are in the same cluster in Y but not in the same cluster in T, whereas a is the number of data point pairs that are neither in the same cluster of Y nor in the same cluster of T.

NMI is defined as follows:

$$NMI(T, Y) = \frac{\sum_{t \in T} \sum_{y \in Y} p(t, y) ln\left(\frac{p(t,y)}{p(t)p(y)}\right)}{\sqrt{H(T) \cdot H(Y)}}, \tag{19}$$

Here, $H(T)$ and $H(Y)$ represent the information entropy of the tags T and Y, respectively. $p(t)$ and $p(y)$ are the marginal distribution of t and y, $p(t, y)$ represents the joint distribution function of t and y.

3.2 scRNA-seq Datasets

In this paper, nine different scRNA-seq datasets were used to do the relevant experimental analysis. The datasets involved in the experiment include Treutlein [22], Ting [23], Pollen [24], Deng [25], Goolam [26], Kolod [27], mECS, Engel4 [28] and Darmanis [29]. The detailed information of the nine scRNA-seq data sets are shown in Table 1.

3.3 Parameters Setting

In this part, we specifically discuss the influence of different parameters with regard to the effectiveness of THSLRR method. We make use of the grid search method to determine

Table 1. The scRNA-seq data sets used in experiments.

Data set	Cells	Genes	Cell type	Species
Treutlein	80	959	5	Homo sapiens
Ting	114	14405	5	Mus musculus
Deng	135	12548	7	Mus musculus
Pollen	249	14805	11	Homo sapiens
Goolam	124	40315	5	Mus musculus
Kolod	704	10685	3	Mus musculus
mECS	182	8989	3	Mus musculus
Engel4	203	23337	4	Homo sapiens
Darmanis	420	22085	8	Homo sapiens

Table 2. The optimal values of four parameters for scRNA-seq data sets.

Data sets	λ_1	λ_2	β	γ
Treutlein	$10^{0.5}$	10^1	10^{-1}	10^1
Ting	10^1	10^1	10^{-4}	10^1
Deng	$10^{0.7}$	$10^{1.1}$	10^{-2}	$10^{-2.3}$
Pollen	10^0	10^1	10^3	$10^{-1.5}$
Goolam	$10^{0.1}$	10^1	10^{-2}	$10^{-1.1}$
Kolod	$10^{0.5}$	10^1	10^{-2}	$10^{-1.1}$
mECS	$10^{0.9}$	10^2	10^2	$10^{-2.2}$
Engel4	$10^{0.8}$	10^{-1}	$10^{2.6}$	$10^{-1.2}$
Darmanis	$10^{0.2}$	10^1	$10^{-1.9}$	$10^{-0.2}$

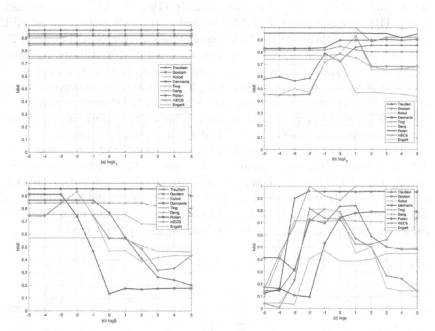

Fig. 2. Sensitivity of different parameters to clustering performance of nine scRNA-seq data sets. (a) λ_1 varying. (b) λ_2 varying. (c) β varying. (d) γ varying.

the optimal combination of parameters. The four parameters change in separate intervals $[10^{-5}, 10^5]$, and when one of the parameters changes, the other parameters are fixed, and then we get Fig. 2. In Fig. 2, the clustering results are insensitive to different λ_1, while λ_2, β and γ have a greater impact on the model performance. Fortunately, within a certain range, we can choose the appropriate combination of parameters to achieve the optimal clustering result. Therefore, we obtain the optimal parameters of different datasets, as shown in Table 2.

3.4 Comparative Analysis of Clustering

We conduct experiments on nine scRNA-seq data sets recounted in Table 1 to discuss the clustering performance of THSLRR. t-SNE, K-means, SIMLR, SC, Corr, MPSSC and SinNLRR are selected as comparison methods. In order to ensure the fairness and objectivity of the comparison, we furnish the real number of classes to THSLRR as well as the other seven methods, and their parameters are all set to the optimal parameters. The comparison results are shown in Fig. 3 and Table 3.

By observing Fig. 3 and Table 3, we can draw the following conclusions:

1) In Fig. 3(a), the median ARI for comparison methods in all datasets is below 0.7, while the median value of THSLRR is greater than 0.9. Furthermore, it is the flattest compared to the box plots of the other seven methods, indicating that the performance of THSLRR is more stable. Similar results can be found in Fig. 3(b).

2) In Table 3, SinNLRR outperforms SIMLR, MPSSC, and Corr on most datasets, and the average ARI for SinNLRR is approximately 11%, 6% and 20% higher, respectively. THSLRR exceeds SIMLR, MPSSC and Corr on all datasets except mECS, and outperforms SIMLR, MPSSC and Corr in terms of average ARI by about 27%, 22% and 36% respectively. As can be seen, the low-rank based clustering methods SinNLRR and THSLRR achieve satisfactory clustering results on most of the data sets, indicating the critical contribution of global information to improve the clustering performance once again. In contrast, SIMLR, MPSSC and Corr only take into consideration the local information between samples, their clustering performance is not as impressive as SinNLRR and THSLRR on most of the datasets.

3) It can also be seen from Table 3 that THSLRR exceeds the SinNLRR method by about 16% in ARI score. There are two main factors. First, the THSLRR method utilizes the hypergraph regularization to thoroughly mine the complex high-order relationships of scRNA-seq data, while sinNLRR simply considers the overall information of the data. Secondly, the similarity based on TRW captures the global manifold structure information of the data and improves the learning ability of the model.

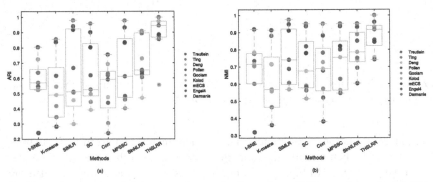

Fig. 3. Clustering results of eight clustering methods on nine scRNA-seq data sets. (a) ARI. (b) NMI

In conclusion, THSLRR achieves the best results on most data sets. Moreover, the average ARI and NMI of THSLRR increase by approximately 12% and 22% compared with comparison methods. Therefore, the THSLRR method is rational and it has certain advantages in cell type identification.

Table 3. The clustering performance on the scRNA-seq data

Method	ARI							
	t-SNE	K-means	SIMLR	SC	Corr	MPSSC	SinNLRR	THSLRR
Treutlein	0.5473	0.6172	0.5114	0.6191	0.5919	0.6117	0.6419	**0.8722**
Ting	0.6384	0.8567	0.9803	0.9592	0.6302	0.9784	0.8943	**1.0000**
Deng	0.5301	0.4914	0.4565	0.3917	0.4753	0.4783	0.4706	**0.5553**
Pollen	0.8055	0.8378	0.9415	0.9013	0.7553	0.9328	0.9051	**0.9448**
Goolam	0.5255	0.4182	0.2991	0.4445	0.3046	0.402	0.9097	**0.9727**
Kolod	0.7265	0.5462	0.2991	0.4974	0.6928	0.8306	0.7291	**0.9727**
mECS	0.2408	0.2824	**0.9186**	0.8028	0.2385	0.8347	0.6263	0.8857
Engel4	0.5725	0.3453	0.6682	0.5258	0.4377	0.4821	0.6533	**0.8554**
Darmnis	0.5725	0.3453	0.5069	0.5258	0.6183	0.4593	0.6057	**0.9452**
average	0.5732	0.5494	0.6202	0.6427	0.5269	0.6678	0.7288	**0.8893**

3.5 Visualize Cells Using t-SNE

According to [6], we make use of the improved t-SNE to map the learned matrix H to the two-dimensional space to observe the structure representation performance of THSLRR method. We only analyze the visualization results for the Ting and Darmanis datasets because of space limitations.

As shown in Fig. 4(a), THSLRR does not distinguish class 1 from class 4 on the Treutlein data, but the boundaries among other types of cells are more obvious. SinNLRR does not distinguish the three cell types 1, 3 and 4, the boundary between classes 2 and 5 is also very blurred. The distribution of t-SNE, SIMLR and MPSSC cells are also scattered. In Fig. 4(b), the result of t-SNE is the worst, SIMLR divides cells belonging to the same class into two clusters, SinNLRR and MPSSC fail to separate the two types of cells and THSLRR can correctly separate five cell types. All methods do not show promising results on the Pollen and Darmanis datasets in Fig. 4(c) and Fig. 4(d), while THSLRR performed best overall because almost all cells belonging to the same cluster are segregated into the same group and the boundaries between clusters were relatively clear.

3.6 Gene Markers Prioritization

In this section, the affinity matrix H learned from THSLRR is used to prioritize genes. First, the bootstrap Laplacian score that is proposed in [6] is used for identifying gene

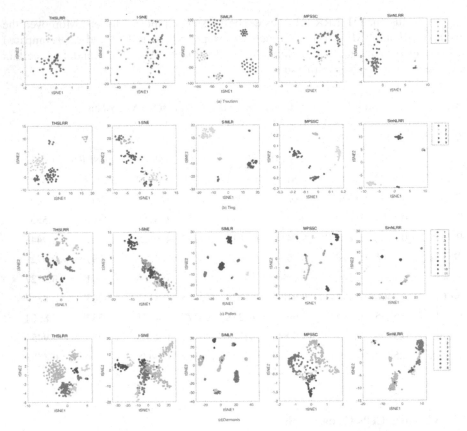

Fig. 4. Visualization results of the cells on (a) Treutlein, (b) Ting, (c) Pollen, and (d) Darmanis datasets.

markers on the matrix H. Then, the genes are placed in descending order in the light of their importance in distinguishing cell subpopulations. Finally, the top ten genes are selected for visual analysis. We use Engel4 and Darmanis data sets for gene markers analysis.

On Darmanis and Engel4 data sets, we select the top 10 gene markers as shown in Fig. 5(a) and Fig. 5(b) respectively. The color of the ring indicates the mean expression level of the gene, and the darker the color, the higher the average expression level of the gene. The size of the ring means the percentage of gene expression in the cell.

Figure 5(a) shows the top ten genes of Darmanis data set. The genes SLC1A3, SLC1A2, SPARCL1 and AQP4 have a high level of expression in astrocytes, and they play an essential part in early development of astrocytes. In fetal quiescent, SOX4, SOX11, TUBA1A and MAP1B have a high level of expression and have been proven to be marker genes with specific roles [30–33]. MAP1B in neurons is also highly expressed PLP12 and CLDND1 with high expression in oligodendrocytes can be regarded as gene markers of oligodendrocytes [34]. In the Engel4 data, as shown in Fig. 5(b), Engel et al. have been confirmed for Serpinb1a, Tmsb10, Hmgb2 and Malta1 [28]. The remaining genes have also been selected as marker genes in related literature [35, 36].

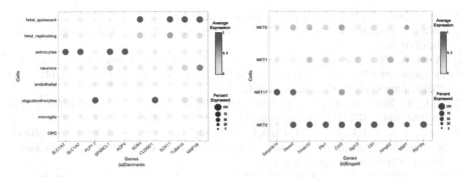

Fig. 5. The top ten gene markers. (a) Darmanis data set. (b) Engel4 data set.

4 Conclusion

In this paper, we propose a clustering method based on subspace learning, named THSLRR. There are mainly two differences where our method differs from other subspace clustering methods. The first aspect is the introduction of hypergraph regularization, which is used to encode higher-order geometric relationships among data and to mine the internal information of data. Compared with other subspace clustering methods, the complex relationships of data can be extracted by our method. Another aspect is the similarity constraint based on TRW, it can mine the global nonlinear manifold structure information of the data and improve the clustering performance and the interpretability of the model. Comparative experiments prove the effectiveness of the THSLRR method. Moreover, the THSLRR method can also provide guidance for data mining as well as be employed in other related domains.

Now, we would like to discuss the limitations of our model. Primarily, although the optimal combination of parameters can be searched by the grid search method, it would be helpful if the optimal parameters could be determined automatically based on some strategy. Second, we use the single similarity criterion in our model, which may not be comprehensive for capturing similarity information from the data. So we can try to use measurement fusion to capture more accurate prior information in the next work.

Funding. This work was supported in part by the National Science Foundation of China under Grant Nos. 62172253 , 61972226 and 62172254.

References

1. Kalisky, T., Quake, S.R.: Single-cell genomics. Nat. Methods **8**(4), 311–314 (2011). https://doi.org/10.1038/nmeth0411-311
2. Pelkmans, L.: Using cell-to-cell variability—a new era in molecular biology. Science **336**(6080), 425 (2012). https://doi.org/10.1126/science.1222161
3. Forgy, E.W.: Cluster analysis of multivariate data: efficiency versus interpretability of classifications. Biometrics **21**(3) (1965). https://doi.org/10.1080/00207239208710779
4. Luxburg, U.V.: A tutorial on spectral clustering. Stat. Comput. **17**(4), 395–416 (2004). https://doi.org/10.1007/s11222-007-9033-z

5. Xu, C., Su, Z.: Identification of cell types from single-cell transcriptomes using a novel clustering method. Bioinformatics **12**, 1974–1980 (2015). https://doi.org/10.1093/bioinformatics/btv088

6. Wang, B., Zhu, J., Pierson, E., Ramazzotti, D., Batzoglou, S.: Visualization and analysis of single-cell RNA-seq data by kernel-based similarity learning. Nat. Methods **14**(4), 414 (2017). https://doi.org/10.1038/nmeth.4207

7. Park, S., Zhao, H., Birol, I.: Spectral clustering based on learning similarity matrix. Bioinformatics **34**(12) (2018). https://doi.org/10.1093/bioinformatics/bty050

8. Jiang, H., Sohn, L.L., Huang, H., Chen, L.: Single cell clustering based on cell-pair differentiability correlation and variance analysis. Bioinform. (Oxf. Engl.) **21**, 3684 (2018). https://doi.org/10.1093/bioinformatics/bty390

9. Zhang, W., Li, Y., Zou, X.: SCCLRR: a robust computational method for accurate clustering single cell RNA-seq data. IEEE J. Biomed. Health Inform. **25**(1), 247–256 (2020). https://doi.org/10.1109/JBHI.2020.2991172

10. Zheng, R., Li, M., Liang, Z., Wu, F.X., Pan, Y., Wang, J.: SinNLRR: a robust subspace clustering method for cell type detection by nonnegative and low rank representation. Bioinformatics (2019). https://doi.org/10.1093/bioinformatics/btz139

11. Liu, G., Yan, S.: Latent Low-Rank Representation for subspace segmentation and feature extraction. IEEE (2012). https://doi.org/10.1109/ICCV.2011.6126422

12. Zheng, R., Liang, Z., Chen, X., Tian, Y., Cao, C., Li, M.: An adaptive sparse subspace clustering for cell type identification. Front. Genet. **11**, 407 (2020). https://doi.org/10.3389/fgene.2020.00407

13. Van der Maaten, L., Hinton, G.: Visualizing data using t-SNE. J. Mach. Learn. Res. **9**(11), 2579–2605 (2008)

14. Liu, G., Lin, Z., Yan, S., Sun, J., Yu, Y., Ma, Y.: Robust recovery of subspace structures by low-rank representation. IEEE Trans. Pattern Anal. Mach. Intell. **35**(1), 171–184 (2012). https://doi.org/10.1109/TPAMI.2012.88

15. Tu, E., Cao, L., Yang, J., Kasabov, N.: A novel graph-based k-means for nonlinear manifold clustering and representative selection. Neurocomputing **143**, 109–122 (2014). https://doi.org/10.1016/j.neucom.2014.05.067

16. Wang, H., Wu, J., Yuan, S., Chen, J.: On characterizing scale effect of Chinese mutual funds via text mining. Signal Process. **124**, 266–278 (2016). https://doi.org/10.1016/j.sigpro.2015.05.018

17. Et, A., Yz, B., Lin, Z.C., Jie, Y.D., Nk, E.: A graph-based semi-supervised k nearest-neighbor method for nonlinear manifold distributed data classification. Inf. Sci. **367–368**, 673–688 (2016). https://doi.org/10.1016/j.ins.2016.07.016

18. Lin, Z., Liu, R., Su, Z.: Linearized alternating direction method with adaptive penalty for low-rank representation. In: Advances in Neural Information Processing Systems, pp. 612–620 (2011). https://doi.org/10.48550/arXiv.1109.0367

19. Chen, J., Mao, H., Sang, Y., Yi, Z.: Subspace clustering using a symmetric low-rank representation. Knowl. Based Syst. **127**, 46–57 (2017). https://doi.org/10.1016/j.knosys.2017.02.031

20. Meilă, M.: Comparing clusterings—an information based distance. J. Multivar. Anal. **98**(5), 873–895 (2007). https://doi.org/10.1016/j.jmva.2006.11.013

21. Strehl, A., Ghosh, J.: Cluster ensembles - a knowledge reuse framework for combining multiple partitions. J. Mach. Learn. Res. 3(3), 583–617 (2002). https://doi.org/10.1162/153244303321897735

22. Treutlein, B., et al.: Reconstructing lineage hierarchies of the distal lung epithelium using single-cell RNA-seq. Nature (2014). https://doi.org/10.1038/nature13173

23. Ting, D.T., et al.: Single-cell RNA sequencing identifies extracellular matrix gene expression by pancreatic circulating tumor cells. Cell Rep. **8**(6), 1905–1918 (2014). https://doi.org/10.1016/j.celrep.2014.08.029

24. Pollen, A.A., et al.: Low-coverage single-cell mRNA sequencing reveals cellular heterogeneity and activated signaling pathways in developing cerebral cortex. Nat. Biotechnol. **32**(10), 1053–1058 (2014). https://doi.org/10.1038/nbt.2967

25. De Ng, Q., Ramskld, D., Reinius, B., Sandberg, R.: Single-Cell RNA-Seq reveals dynamic, random monoallelic gene expression in mammalian cells. Science **343** (2014). https://doi.org/10.1126/science.1245316

26. Goolam, M., et al.: Heterogeneity in Oct4 and Sox2 targets biases cell fate in 4-cell mouse embryos. Cell **165**(1), 61–74 (2016). https://doi.org/10.1016/j.cell.2016.01.047

27. Kolodziejczyk, A.A., et al.: Single cell RNA-sequencing of pluripotent states unlocks modular transcriptional variation. Cell Stem Cell **17**(4), 471–485 (2015). https://doi.org/10.1016/j.stem.2015.09.011

28. Engel, I., et al.: Innate-like functions of natural killer T cell subsets result from highly divergent gene programs. Nat. Immunol. (2016). https://doi.org/10.1038/ni.3437

29. Darmanis, S., et al.: A survey of human brain transcriptome diversity at the single cell level. Proc. Natl. Acad. Sci. U. S. A. **112**(23), 7285–7290 (2015). https://doi.org/10.1073/pnas.1507125112

30. Takemura, R., Okabe, S., Umeyama, T., Kanai, Y., Hirokawa, N.: Increased microtubule stability and alpha tubulin acetylation in cells transfected with microtubule-associated proteins MAP1B, MAP2 or tau. J. Cell Sci. **103**(Pt 4), 953–964 (1993). https://doi.org/10.1083/jcb.119.6.1721

31. Uwanogho, D., et al.: Embryonic expression of the chicken Sox2, Sox3 and Sox11 genes suggests an interactive role in neuronal development. Mech. Dev. **49**(1–2), 23–36 (1995). https://doi.org/10.1016/0925-4773(94)00299-3

32. Medina, P.P., et al.: The SRY-HMG box gene, SOX4, is a target of gene amplification at chromosome 6p in lung cancer. Huma. Mol. Genet. **18**(7), 1343 (2009). https://doi.org/10.1093/hmg/ddp034

33. Cushion, T.D., et al.: Overlapping cortical malformations and mutations in TUBB2B and TUBA1A. Brain A J. Neurol. **2**, 536–548 (2013). https://doi.org/10.1093/brain/aws338

34. Numasawa-Kuroiwa, Y., et al.: Involvement of ER stress in dysmyelination of pelizaeus-merzbacher disease with PLP1 missense mutations shown by iPSC-derived oligodendrocytes. Stem Cell Rep. **2**(5), 648–661 (2014). https://doi.org/10.1016/j.stemcr.2014.03.007

35. Yu, N., Liu, J.X., Gao, Y.L., Zheng, C.H., Shang, J., Cai, H.: CNLLRR: a novel low-rank representation method for single-cell RNA-seq data analysis. Hum. Genomics (2019). https://doi.org/10.1101/818062

36. Jiao, C.-N., Liu, J.-X., Wang, J., Shang, J., Zheng, C.-H.: Visualization and analysis of single cell RNA-seq data by maximizing correntropy based non-negative low rank representation. IEEE J. Biomed. Health Inform. (2021). https://doi.org/10.1109/JBHI.2021.3110766

Traceability Analysis of Feng-Flavour Daqu in China

Yongli Zhang[1], Chen Xu[2], Gang Xing[2], Zongke Yan[2], and Yaodong Chen[1(✉)]

[1] College of Life Sciences, Northwestern University, Xi'an 710069, Shaanxi, China
ydchen@nwu.edu.cn
[2] Shaanxi Xifeng Liquor Co., Ltd., Baoji 721400, Shaanxi, China

Abstract. High throughput sequencing was used to analyze the microbial community landscape of Chinese Feng-Flavour Daqu, and to study the specific contribution of different environmental factors to Daqu microorganisms. Taking the microbial population of the raw materials (wheat, pea and barley) and the environmental samples (tools, indoor ground, outdoor ground and air) as the source, and the microbial population of Feng-Flavour Daqu as the receiver, software Source-Tracker was used to trace and analyze the microorganisms in Feng-Flavour Daqu. 94.7% of the fungi in the newly pressed Feng-Flavour Daqu come from raw materials, 1.8% from outdoor ground and 3.47% from unknown environment; 60.95% of bacteria come from indoor ground, 20.44% from raw materials, 8.98% from tools, and the rest from unknown environment. The source of main microorganisms in Feng-Flavour Daqu and the influence of environmental factors on the quality of Daqu were clarified, which provided a basis for improving the quality of Feng-Flavour Daqu.

Keywords: Feng-Flavour Daqu · Microorganism · Environment · Traceability

1 Introduction

The flavor is the style characteristics of liquor, which is used to distinguish the difference of the characteristic liquor in China. At present, there are 12 liquor flavor types in China and each one has its unique flavoring characteristics. Different flavor types are mainly due to different production areas and processes. Feng-Flavour Liquor is one of the four famous traditional liquors in China. It has the characteristics of elegant liquor flavor, enjoyable liquor taste, harmonious liquor body and long liquor aftertaste.

Daqu, an undefined starter culture, is one kind of Jiuqu (a sort of equivalence of Koji) [1]. Daqu is commonly known as "Bone of liquor". Daqu contains a variety of microorganisms used for the fermentation of Chinese liquor. Among them, fungi dominated by molds and yeasts are an important functional microorganism, which can secrete amylase, cellulase and other enzymes [2]. The microbes of the Daqu are one of the determinants of the style and taste of liquor. The production process of Feng-Flavour Daqu is divided

© The Author(s), under exclusive license to Springer Nature Switzerland AG 2022
H. Han and E. Baker (Eds.): SDSC 2022, CCIS 1725, pp. 94–106, 2022.
https://doi.org/10.1007/978-3-031-23387-6_7

into four stages: crushing of raw materials, pressing and forming of raw materials containing certain moisture by machine, placing the formed Daqu into the culture room for 30 days, and storing it in the warehouse for 3 months. The production of Daqu is a spontaneous solid-state fermentation process with natural inoculation [3]. Raw materials without high-pressure sterilization will be exposed to many environments (such as air, ground, etc.) during the stage from raw materials to mature Daqu [4]. Therefore, microorganisms in the environment, especially in the liquor production area, are also one of the important sources of Daqu flora [5]. Environmental microorganisms in specific areas may be one of the reasons for the specific flavor and types of Chinese liquor in different regions.

Microbial traceability originated in the last century and is mostly used in the study of water pollution to identify pollution sources [6, 7]. At present, microbial traceability analysis has also been widely used in the analysis of microorganisms in soil [8], air [9], coral [10]. Further, the idea of microbial traceability analysis promotes the monitoring of microbial sources in the field of food fermentation. For example, Stellato et al. detected the distribution of microorganisms related to meat product corruption in the environment of meat products processing plant (knife, chopping board and workers' hands) through 16SrRNA amplicon sequencing technology, and detected more than 800 OTUs in the food processing environment, indicating that the microbial composition of food processing environment was complex, and most environmental microorganisms could be detected in meat products [11]. Doyle et al. detected the microbial population distribution in the indoor and outdoor environment of raw milk collection in different seasons through amplicon sequencing technology, the results showed that the microorganisms in the raw milk collected indoors and outdoors were related to the environment (grass, feed, feces, soil and milker) [12]. At the same time, they believed that the milker was the main source of microorganisms in the raw milk. The environmental microecology test of cheese factory also found that the microorganisms on the surface of processing equipment also participated in the fermentation process, due to the differences of cheese types and maturity, the microorganisms on the surface of equipment also formed different microecology [13]. The micro ecology of food fermentation environment is also affected by geographical factors, which have been reported in many foods production. Bokulich et al. discovered that the microbial population in wine grape was related to grape varieties, harvest areas and climate environment, and the microorganisms in grape planting soil and grapes showed the same regional distribution, indicating that soil microorganisms were an important source of wine grape microorganisms [14]. Therefore, different grape planting areas formed different microbial groups in grapes, thus forming a unique wine style in different regions. Knight et al. further proved that the landmark characteristics of wine were related to microorganisms by using different characteristics of Saccharomyces cerevisiae to ferment everlasting longing for each other grapes [15]. The landmark characteristics of this fermented food not only appear in wine. Bokulich et al. found that the type of milk and the origin of milk would affect the microbial population structure in fermented dairy products, while the traditional workshop production method ensured that the microbial characteristics in different regions could be inherited from generation to generation [16].

SourceTracker is a tool for quantitative analysis of microbial sources based on Bayesian reasoning [17]. Different from the traditional identification of indicator microorganisms, SourceTracker finds out the source of target microorganisms by comparing the similarity of microbial community structure between samples and pollution sources, and its accuracy is much higher than that of traditional random forest method and Naive Bayes model. The migration direction of microorganisms or genes in the environment can be monitored through SourceTracker, which is widely used in many research areas. Bokulich et al. used SourceTracker to analyze the distribution and migration of microorganisms related to pollution in breweries [14]. Doyle et al. combined with amplicon sequencing technology and source tracker analysis, showed that the microorganisms in raw milk collected indoors and outdoors were related to the environment (grass, feed, feces, soil and milker), and considered that milker was the main source of microorganisms in raw milk [12]. Du et al. used SourceTracker to explore the impact of raw materials and environment on the microbiota of Chinese Daqu, they found that the fungal community in new Daqu mainly comes from the Daqu production environment (mainly tools and indoor ground), most of the bacterial community in Daqu comes from raw materials [18]. Zhou et al. analyzed the source of microorganisms in Gujing tribute liquor Daqu through SourceTracker and found that the bacteria in Daqu at the beginning of fermentation mainly came from raw materials and the fungi came from outdoor ground [5].

This open fermentation of Chinese liquor has brought about beneficial microorganisms fermented by the environment, but also enriched a number of useless or harmful microorganisms. This brings challenges to the quality and safety in production. The batch instability of traditional fermented food also comes from the uncontrollability of environmental microorganisms. Therefore, the traceability of fermented food microorganisms, especially the traceability of environmental microorganisms, is very important to control and improve the quality of liquor.

Different raw materials and processes of Daqu not only affect the source of microorganisms, but also create their unique styles and characteristics. Therefore, we took raw materials (wheat, pea and barley) and environmental samples (tools, indoor ground, outdoor ground, water and air) as the source of microorganisms in Feng-Flavour Daqu, and analyzed them to understand their unique characteristics. In this study, we clarified the influence of environmental factors on microbial changes during the maturation of Feng-Flavour Daqu, which is the basis for further improving the quality of Daqu.

2 Materials and Methods

2.1 Sample Collection

Nine types of samples were collected, including Feng-Flavour Daqu, raw materials for Daqu production (wheat, pea and barley) and environmental samples (tools for Daqu production, indoor ground, outdoor ground and air), as follows (Table 1):

Table 1. Samples in this study.

Sample	Description
Newly pressed Daqu (NDaqu)	Daqu that has just been pressed by machine and has not yet been fermented in the room
Mature Daqu (Daqu)	A Daqu which can be used for Feng-Flavour liquor production after 3 months' storage
Raw materials (RM)	Raw materials from the raw material crushing workshop
Enhanced strains (EDS)	The mixed fortified strains
Tools	The machine for pressing Daqu, the cart for transporting Daqu, the bamboo mat, bamboo and rice bran in contact with Daqu
Indoor ground (ING)	The door, window and middle floor of the room where Daqu is cultivated
Outdoor ground (OUTG)	The sidewalk outside the room where Daqu is cultivated
Air	Air in workshop
Water	Water for Daqu production

For Newly pressed Daqu (NDaqu), Mature Daqu (Daqu), Raw materials (RM) and Enhanced strains (EDS), randomly select three places, and take 50 g of each place as a sample. For tools, Indoor ground (ING) and Outdoor ground (OUTG), randomly select three points, wipe the surface with sterile cotton soaked in $0.1 \ mol \cdot L^{-1}$ PBS buffer, and put the sample into a sterile self-sealing bag. For Air, before sampling, the collector and catheter are ultrasonically cleaned and dried, and then 20 mL of $0.1 \ mol \cdot L^{-1}$ PBS buffer is filled into the collector. Place the sampler 2 m above the ground and collect at a speed of $10 \ L \cdot min^{-1}$ for 2.5 h. The collected PBS solution was filtered with $0.22 \ \mu M$ filter membrane to collect air microbial samples. For Water, take three portions of water for Daqu production, 2000 mL each, and filter them with $0.22 \ \mu M$ filter membrane to collect microbial samples in the water.

2.2 Microbial High-Throughput Sequencing Analysis of Traceable Samples

Extraction of total DNA from microbiota with DNA kit (Omega Bio-Tek, USA). Nanodrop were used to quantify DNA and the quality of DNA was determined by 1.2% agarose gel electrophoresis. The extracted DNA from samples were stored at $-80°C$ for amplicon sequencing.

Amplicon sequencing: The fungal sequencing region was ITS_V1, using primers ITS5F (GGAAGTAAAAGTCGTAACAAGG) and ITS2R (GCTGCGTTCTTCATC-GATGC).

The bacterial sequencing region was 16S v3–v4, using primers F (ACTCC-TACGGGAGGCAGCA) and R (GGACTACHVGGGTWTCTAAT). After amplification, the purified 16S rRNA gene and ITS1 sequences were sequenced by Illlumina MiSeq platform, respectively, at BioNovoGene Co., Ltd. (Suzhou, China).

Sequence analysis: Microbiome bioinformatics were performed with QIIME2 2019.4 [19] with slight modification according to the official tutorials. Briefly, raw sequence data were demultiplexed using the demux plugin following by primers cutting with cutadapt plugin. Sequences were then quality filtered, denoised, merged and chimera removed using the DADA2 plugin. Non-singleton amplicon sequence variants (ASVs) were aligned with mafft and used to construct a phylogeny with fasttree2. Alpha-diversity metrics (Chao1, Observed species, Shannon, Simpson, Faith's PD, Pielou's evenness and Good's coverage), beta diversity metrics (weighted UniFrac, unweighted UniFrac, Jaccard distance, and Bray-Curtis dissimilarity) were estimated using the diversity plugin with samples were rarefied to 16166 (bacterial) and 2018 (fungal) sequences per sample. Taxonomy was assigned to ASVs using the classify-sklearn naïve Bayes taxonomy classifier in feature-classifier plugin against the SILVA Release 132/UNITE Release 8.0 Database.

Bioinformatics and statistical analysis: Sequence data analyses were mainly performed using QIIME2 and R packages (v3.2.0). ASV-level alpha diversity indices, such as Chao1 richness estimator, Observed species, Shannon diversity index, Simpson index, Faith's PD, Pielou's evenness and Good's coverage were calculated using the ASV table in QIIME2, and visualized as box plots. ASV-level ranked abundance curves were generated to compare the richness and evenness of ASVs among samples. Beta diversity analysis was performed to investigate the structural variation of microbial communities across samples using Jaccard metrics, Bray-Curtis metrics and UniFrac distance metrics and visualized via nonmetric multidimensional scaling (NMDS) and unweighted pair-group method with arithmetic means (UPGMA) hierarchical clustering. The taxonomy compositions and abundances were visualized using MEGAN [20] and GraPhlAn [21].

2.3 Traceability Analysis of Brewing Microorganisms

Venn diagram can directly reflect the common and unique microbial populations of Daqu and the environment. Venn diagram was generated to visualize the shared and unique ASVs among samples or groups using R package "VennDiagram", based on the occurrence of ASVs across samples/groups regardless of their relative abundance. In this study, each type of sample contains three parallels. We only select the bacteria that appear in all three parallels as valid data, and then conduct genus level Venn analysis through the screened data.

Traceability analysis of fermentation microorganisms: according to the microbial population structure of Daqu and fermentation environment, this study uses Source-Tracker software to analyze the source of microorganisms in Daqu, and sets the microbial population of Daqu raw materials (wheat, pea and barley) and environmental samples (tools, indoor ground, outdoor ground and air) as the source, the microbial population of Feng-Flavour Daqu as the receiving end, running 1000 times, and other parameters are default.

3 Results

3.1 Analysis of Fungal Community Diversity in Feng-Flavour Daqu and Its Environment

The dilution curve shows that all samples have reached the platform stage, indicating that this sequencing can cover the vast majority of fungal population information in the samples (Fig. 1a). New pressed Daqu and raw materials share the highest ASV (194), while mature Daqu and air share the highest ASV (31) (Fig. 1b). The Venn diagram results of ASV of each sample show that the number of unique ASV in the air is the most and the mature Daqu is the least, there are 7 ASVS common to all samples (Fig. 1c).

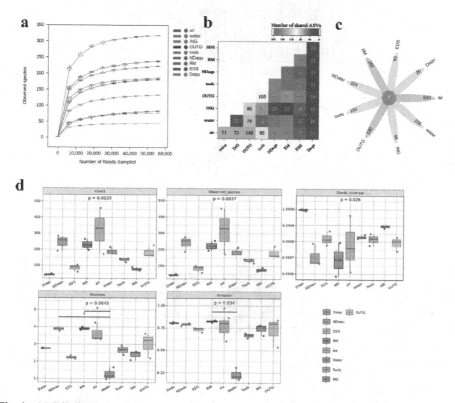

Fig. 1. (a) Dilution curve of fungal population. This sequencing can cover the vast majority of fungal population information in the samples. (b) Amplicon sequence variants (ASVS) distribution of fungi in each sample. NDaqu and RM share the highest ASV, while Daqu and air share the highest ASV. (c) Venn diagram is used to represent the ASV shared by fungi in the experimental sample, there are 7 ASVS common to all samples. (d) The fungal population diversities of Daqu, RM and environment were evaluated by richness index (Chao1 and observed species), good's coverage index (good's coverage) and diversity index (Shannon and Simpson). The richness and diversity of fungi is the highest in the air and the lowest in Daqu. Note: For the sake of concise expression, in to ground, out ground, new Daqu, raw materials and enhanced trains in this article are abbreviated as ING, OUTG, NDaqu, RM and EDS respectively.

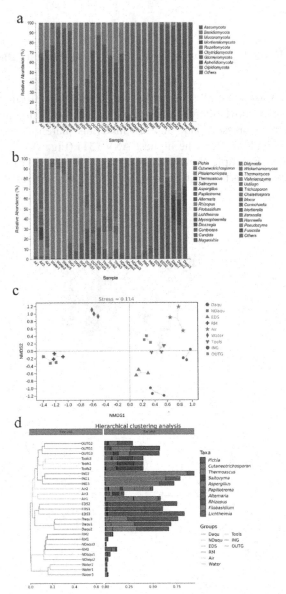

Fig. 2. (a) Analysis of fungal population structure of each sample at phylum level, 9 fungal phyla were detected, the dominant fungi phyla were *Ascomycota* and *Basidiomycota*. (b) Analysis of fungal population structure of each sample at genus level, 301 fungal genera were detected, the dominant fungi in different samples were vary greatly. (c) Nonmetric multidimensional scaling analysis (NMDS) of fungal population suggests that NDaqu and RM are close, and Daqu and EDS are close. (d) Hierarchical cluster analysis (HCA) of fungal population of different samples.

In this study, the fungal population diversity of Daqu, raw materials and environment was evaluated by richness index (Chao1 and observed species), good's coverage index (good's coverage) and diversity index (Shannon and Simpson) (Fig. 1d). The results of diversity analysis showed that the fungal richness of mature Daqu was lower than that of newly pressed Daqu, and the fungal richness of mature Daqu was the lowest in all samples. In environmental samples, the indoor ground soil fungal richness was the lowest and the air was the highest. The sequencing depth of all samples in the figure is greater than 0.9995, indicating that the sequencing depth has basically covered all species in the samples. According to Shannon index and Simpson index, the fungal diversity of mature Daqu is lower than that of newly pressed Daqu, and the fungal diversity of raw materials is basically the same as that of newly pressed Daqu. In environmental samples, the fungal diversity in air is the highest and in water is the lowest.

A total of 301 genera were detected at the genus level, of which 85 genera could be detected in newly pressed Daqu, only 33 genera in mature Daqu, and 202 genera could only be detected in raw materials or environmental samples (Fig. 2). The relative abundances of three ASVs in the new pressed Daqu were > 1%, and the relative abundances of the top two were 63.58 ± 6.11% and 31.51 ± 7.33% respectively, but the two ASVs were not annotated to the genus level, and *Pichia* ranked third. The dominant genera in mature Daqu are *Pichia, Thermoascus* and *Aspergillus*.

In order to further explain the relationship between Daqu raw materials, environment and Daqu microorganisms, the nonmetric multidimensional scaling (NMDS) and hierarchical clustering analysis (HCA) based on Jaccard distance were applied to analyze the data. The analysis results of NMDS and HCA (Fig. 2c–d) show that the newly pressed Daqu is closest to the raw materials, the mature Daqu is closest to the enhanced strains, the water and other samples are far away, and the air, tools and ground samples are close.

3.2 Diversity Analysis of Bacteria and Environmental Communities in Feng-Flavour Daqu

The dilution curve shows that all samples have reached the platform stage, indicating that this sequencing can cover the vast majority of bacterial population information in the samples (Fig. 3a). The number of new pressed Daqu and fortified strains was the highest (283), followed by raw materials (200) and tools (214). The number of ASVs shared by mature Daqu and tools is the highest (86), followed by indoor ground (75) (Fig. 3b). The results of ASV Wayne diagram of each sample show that the unique ASV in the tool is the most (3127), and the unique ASV in the newly pressed Daqu is the least, only 341, the number of ASV detected in all samples is 0 (Fig. 3C).

The results of bacterial diversity analysis showed that the bacterial richness of mature Daqu was higher than that of newly pressed Daqu, and the bacterial richness of raw materials was the lowest in all tested samples; Except for tools, the sequencing depth of other samples is greater than 0.99, indicating that the sequencing depth has basically covered all species in the samples. The bacterial diversity in mature Daqu is higher than that in newly pressed Daqu. Among all tested samples, the bacterial diversity in raw materials is the lowest, that in tools is the highest, and that in indoor ground is the second (Fig. 3d).

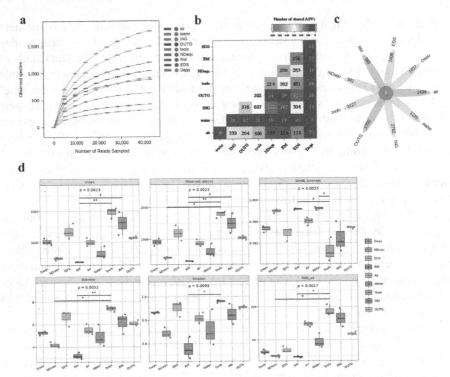

Fig. 3. (a) Bacterial population density curve. This sequencing basically covers the majority of fungal population information in the sample. (b) ASV distribution of bacterial in each sample shows that NDaqu and EDS, Daqu and tools share the highest ASV respectively. (c)Venn diagram is used to represent the AVS shared by bacteria in the experimental sample. Nine samples had no shared ASV. (d) The bacterial population diversity of Daqu, RM and environment was evaluated by richness index (Chao1 and observed species), good's coverage index (good's coverage) and diversity index (Shannon and Simpson). The richness and diversity of fungi in the tools were the highest, and the richness and diversity of fungi in Daqu were higher than those in NDaqu.

A total of 682 genera were detected at the genus level, of which 105 genera could be detected in newly pressed Daqu, only 50 genera in mature Daqu, and 555 genera could only be detected in other samples such as environment (Fig. 4a-b). The dominant bacteria in the newly suppressed Daqu are *Pantoea*, *Chloroplast*, *Leuconostoc* and *Erwinia*. The dominant genera in mature Daqu are *Bacillus*, *Streptomyces*, *Saccharopolyspora*, *Lactobacillus*, *Kroppenstedtia*, *Pseudonocardiaceae*, *Weissella*, *Staphylococcus* and *Acetobacter*.

The results of NMDS analysis and HCA analysis show that the microbial population of newly pressed Daqu is closest to the raw material, while the mature Daqu is far away from other samples (Fig. 4c–d). Therefore, we speculate that raw materials are the main source of bacterial population in newly pressed Daqu. Through the process of Daqu Culture and storage, the bacterial population in newly pressed Daqu gradually succession to the bacterial population structure in mature Daqu.

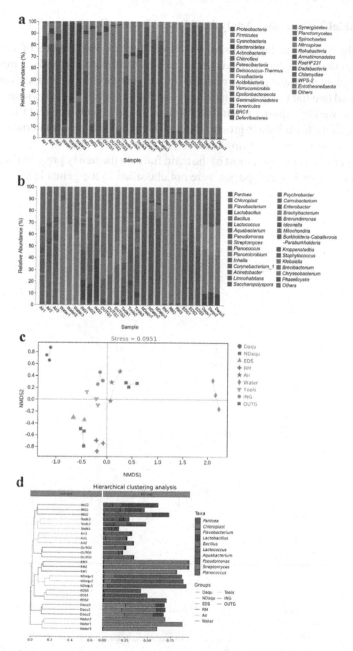

Fig. 4. (a) Analysis of bacterial population structure of each sample at phylum level. 27 bacterial phyla were detected and the dominant bacterial phyla were *Firmicutes, Proteobacteria, Actinobacteria, Bacteroidetes* and *Cyanobacteria*. (b) Analysis of bacterial population structure of each sample at genus level. 682 bacterial genera were detected, and the dominant bacteria in different samples were vary greatly. (c) Nonmetric multidimensional scaling analysis (NMDS) of bacterial population shows that NDaqu and RM, Daqu and ING shares the nearest distance. (d) Hierarchical cluster analysis (HCA) of bacterial population suggests that NDaqu and RM shares the nearest distance.

3.3   Microbial Traceability Analysis of Feng-Flavour Daqu

In this study, SourceTracker software was used to track the sources of microorganisms in Daqu with potential source microorganisms (raw materials and environment) of Daqu as the source end and newly pressed Daqu microorganisms as the receiver end. The results showed that the fungi in the newly pressed Daqu mainly came from raw materials (94.7%), followed by outdoor ground (1.8%) and unknown environment (3.47%); Bacteria mainly came from indoor ground (60.95%), followed by raw materials (20.44%), tools (8.98%) and unknown environment (9.63) (Fig. 5a).

Raw materials contributed most of the main fungi in the newly pressed Daqu, among which the most contributing species were not classified to the genus level. In addition,

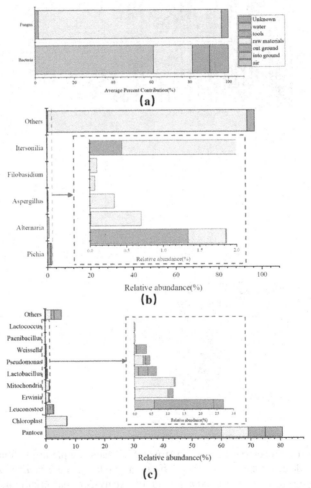

Fig. 5. (a) Microbial traceability analysis of NDaqu to determine the contribution rate of different sources. 94.7% of fungi come from RM and 60.95% of bacteria come from ING. (b) Fungal traceability shows that RM contribute to the most. (c) Bacterial traceability shows that *Pantophytic* are the dominant bacteria, which mainly come from ING.

they also contributed a small amount of *Pichia, Alternaria, Aspergillus, Filobasidium, Itersonia*, etc., and the outdoor ground contributed 1.34% of *Pichia* (Fig. 5b). *Pantophytes* are the dominant bacteria in the newly pressed Daqu, with 60.02% coming from the indoor ground (Fig. 5C).

4 Discussion and Conclusion

The main sources of fungi in the newly pressed Feng-Flavour Daqu were raw materials (94.7%), and bacteria were mainly from indoor ground (60.95%) and raw materials (20.44%).

This is different from the source of microorganisms in Fen-Flavour and Luzhou-flavor Daqu. The main sources of fungi in Fen-Flavour new pressed Daqu are tools (55.18%) and indoor ground (15.97%). At the beginning of Luzhou-Flavor Daqu fermentation, 53.7% of fungi came from outdoor ground and indoor roof contributed 23.0%. The main source of bacteria in Fen-Flavour and Luzhou-Flavor new pressed Daqu is raw materials.

This may be related to the different sampling of new pressed Daqu. Feng-Flavour Daqu is directly taken, just pressed and formed, and the new Daqu that has not entered the room has relatively little contact with tools and environment. Fen-Flavour Daqu and Luzhou-Flavor Daqu are taken from Daqu that has entered the room and has not yet started cultivation, and have been fully contacted with transportation and cultivation tools; The different sources of three kinds of Daqu fungi may also be caused by different raw materials. The raw materials of Fen-Flavour Daqu are barley and pea, the raw materials of Feng-Flavour Daqu are barley, pea and wheat, and the raw materials of Luzhou-Flavor Daqu are wheat. Different raw materials have selectivity for microbial enrichment; Of course, the microbial community structure in different regions is different, which may also lead to the differences of Daqu microorganisms.

The yeast in Daqu plays a role in saccharification, liquefaction and fermentation in wine production, and also plays a certain role in the production of flavor substances, while the bacteria in Daqu are mostly related to the production of flavor substances. Different flavor types of Daqu have different microbial sources. Raw materials and environment determine the quality of Daqu. Specific production areas have specific microbial communities, forming a unique microbial structure in Daqu and giving Daqu a specific flavor type. In addition to these natural factors, the quality of Feng-Flavour Daqu also depends on our artificially cultured enhanced strains. In previous experiments, we also tried to use non enhanced strains, but the quality is poor. Therefore, the quality of enhanced strains is also one of the key factors affecting the quality of Feng-Flavour Daqu.

Acknowledgements. This research was funded by the National Natural Science Foundation of China (Grant No. 31970050).

References

1. Zhu, Y., Tramper, J.: Koji - where East meets West in fermentation. Biotechnol. Adv. **31**, 1448–1457 (2013)

2. Wang, B.W., Wu, Q., Xu, Y., Sun, B.G.: Specific volumetric weight-driven shift in microbiota compositions with saccharifying activity change in starter for Chinese Baijiu fermentation. Front Microbiol. **9**, 2349 (2018)

3. Peng, L., et al.: Study on the quality of medium-high temperature Daqu in differentcurved layers. Food Ferment. Ind. **46**, 58–64 (2020)

4. Huang, Y.H., et al.: Metatranscriptomics reveals the functions and enzyme profiles of the microbial community in Chinese Nong-flavor liquor starter. Front Microbiol. **8**, 1747 (2017)

5. Zhou, T., et al.: Exploring the source of microbiota in medium-high temperature Daqu based on high-throughput amplicon sequencing. Food Ferment Ind. **47**, 66–72 (2021)

6. Scott, T.M., Rose, J.B., Jenkins, T.M., Farrah, S.R., Lukasik, J.: Microbial source tracking: Current methodology and future directions. Appl. Environ. Microb. **68**, 5796–5803 (2002)

7. Simpson, J.M., Santo Domingo, J.W., Reasoner, D.J.: Microbial source tracking: state of the science. Environ. Sci. Technol. **36**, 5279–5288 (2002)

8. Sun, R.B., et al.: Fungal community composition in soils subjected to long-term chemical fertilization is most influenced by the type of organic matter. Environ Microbiol. **18**, 5137–5150 (2016)

9. Wilkins, D., Leung, M.H.Y., Lee, P.K.H.: Indoor air bacterial communities in Hong Kong households assemble independently of occupant skin microbiomes. Environ Microbiol. **18**, 1754–1763 (2016)

10. Staley, C., et al.: Differential impacts of land-based sources of pollution on the microbiota of southeast Florida coral reefs. Appl. Environ. Microb. **83**, e03378-e3416 (2017)

11. Stellato, G., La Storia, A., De Filippis, F., Borriello, G., Villani, F., Ercolini, D.: Overlap of spoilage-associated microbiota between meat and the meat processing environment in small-scale and large-scale retail distributions. Appl. Environ. Microb. **82**, 4045–4054 (2016)

12. Doyle, C.J., Gleeson, D., O'Toole, P.W., Cotter, P.D.: Impacts of seasonal housing and teat preparation on raw milk microbiota: a high-throughput sequencing study. Appl. Environ. Microb. **83**, e02694-e2716 (2017)

13. Bokulich, N.A., Mills, D.A.: Facility-specific "house" microbiome drives microbial landscapes of artisan cheesemaking plants. Appl. Environ. Microb. **79**, 5214–5223 (2013)

14. Bokulich, N.A., Thorngate, J.H., Richardson, P.M., Mills, D.A.: Microbial biogeography of wine grapes is conditioned by cultivar, vintage, and climate. Proc. Natl. Acad. Sci. USA **111**, E139–E148 (2014)

15. Knight, S., Klaere, S., Fedrizzi, B., Goddard, M.R.: Regional microbial signatures positively correlate with differential wine phenotypes: evidence for a microbial aspect to terroir. Sci. Rep. UK **5**, 14233 (2015)

16. Bokulich, N.A., Bergsveinson, J., Ziola, B., Mills, D.A.: Mapping microbial ecosystems and spoilage-gene flow in breweries highlights patterns of contamination and resistance. Elife **4**, e04634 (2015)

17. Knights, D., et al.: Bayesian community-wide culture-independent microbial source tracking. Nat. Methods **8**, 761–763 (2011)

18. Du, H., Wang, X.S., Zhang, Y.H., Xu, Y.: Exploring the impacts of raw materials and environments on the microbiota in Chinese Daqu starter. Int. J. Food Microbiol. **297**, 32–40 (2019)

19. Bolyen, E., et al.: Reproducible, interactive, scalable and extensible microbiome data science using QIIME 2. Nat. Biotechnol. **37**, 852–857 (2019)

20. Huson, D.H., Auch, A.F., Qi, J., Schuster, S.C.: MEGAN analysis of metagenomic data. Genome Res. **17**, 377–386 (2007)

21. Graphlan homepage. https://github.com/biobakery/graphlan

Visualization of Functional Assignment of Disease Genes and Mutations

Hisham Al-Mubaid(⊠)

University of Houston-Clear Lake, Houston, TX 77062, USA
hisham@uhcl.edu

Abstract. Visualization is one of the important components of data science. This paper presents a method that utilizes the functional annotations of human genes for identifying, analysing, and visualizing the most important and highly represented biological process functions in disease mutations. The analysis and visualization of human gene functions are important for understanding disease progression, gene-disease associations, and disease-gene-mutation relationships. In the past two decades, a number of research projects have been proposed for the analysis and discovery of gene functions and gene functional annotations. However, little work has been done for the functional analysis and annotation of mutations and genetic variants. Effectively identifying significant genetic functions of disease mutations can benefit medical applications related to genetic treatment of hard diseases like some cancer types. We present experiments and results involving more than 25,000 human genes with more than 220,000 genetic mutations from two of the most commonly used mutation databases. We used heat maps for visualization of the clustered biological process functions from the *Gene Ontology* among the disease mutations.

Keywords: Data visualization · Gene functional visualization · Disease and gene mutation

1 Introduction

In Data Science, visualization of data and results is an important step to understand the problem solution and the data attributes. Data Science is a discipline that relies mainly on data to solve problems, extract knowledge from data, and find important patterns [23–36]. In this paper, we use visualization as a tool to illustrate the importance of certain (*biological*) functions in the context of disease mutations. We will explain the problem and show how data visualization can help in understanding the task and the solution. In genomics, one of the important areas of research, is the discovery and assignment of gene functions; also called gene functional annotation [1–3]. In general, any change in the DNA sequence (of a gene) is called a *mutation* or *genetic variant* [1–5]. Such a change can cause one or more diseases or medical conditions and so it is important to study and understand the mutations [3–5].

Typically, the change in the DNA is called *polymorphism* if it is common in the population and appears frequently [6–8]. Moreover, each mutation can be roughly classified into: *benign* or *pathogenic*; such that benign mutations, also called *neutral*, do not cause diseases while pathogenic mutations are related to diseases. That is, if a mutation affects the functionality of a gene in the negative way then it can cause a disease or medical condition [6, 7, 9]. The most common type of genetic variation is the *Single nucleotide polymorphism* (SNP) and is caused by only one change in one nucleotide in the DNA sequence [6]. Diseases and medical conditions resulting from changes in the DNA are collectively called *genetic disorders* [6, 7]. Among the most active genetic diseases are *sickle cell anemia, homeochromatosis,* and *cystic fibrosis* [6]. In the context of gene mutations and genetic variants, the most common research areas include: – classifying mutations into pathogenic or neutral mutations [2, 4]; – mutation disease associations [9]; – discovering new gene mutations [7, 10]; and – analysis of mutation functions [2, 9, 11, 12]. This paper contributes into the general area of functional genomics and data visualization. The functional genomics field is interested in analyzing and discovering new functions of the various biological entities such as genes. In the general area of functional genomics, we are focusing in this paper in the analysis, exploration, and visualization of mutation functions. Specifically, we would like to identify, highlight and visualize the most important and most highly represented functions among disease mutations. The proposed technique relies primarily on the graphics and visualizations of mutation functions from the *Biological Process* taxonomy in the *Gene Ontology*. We employ the functional gene annotations (*GOA*) from the *Gene Ontology* (GO) to identify important and significant *Biological Process* functions related to mutations. We conducted experiments with large number of mutations obtained from two commonly used mutation databases [1–3]. The main goal is to identify the most represented (*most prominent*) *biological process* functional annotations from the *GO* related to disease mutations. Then, we use data visualization to illustrate the identified (biological process) functions that are important and highly represented, and will be illustrated with heatmaps for effective and accurate visualization of these important functions.

2   Identification and Visualization of Biological Process Functions

In this work, we would like to identify the most important biological process functions that can be associated with one or more mutations. Then, we utilize *heatmap* for visualization and analysis of the most important functions in the context of mutations [16–22]; which will help in identifying the functional consequences of a given mutation [2–5, 13]. A pathogenic mutation is typically associated with some malfunctionality at the gene level in stopping the gene from performing its functions normally. Such a pathogenic mutation may affect one of the gene functions which may include, for example, absence of gene function [3]. Using the Biological Process (bp) taxonomy of the Gene Ontology (GO) [5], we want to find which function is enriched more in the given target mutation, or in the given target set of mutations. For functional annotation of genes and all genetic data, the *GO* is the most widely used as it is the main and most comprehensive resource of functions [12, 14, 15]. The gene ontology consists of three taxonomies: biological process, molecular function, and cellular component. Each one of these taxonomies is a *tree-like* directed acyclic graph (DAG). In the biological process taxonomy, each function is a node in this tree. Our proposed method is based on identifying the most common *bp* functions for the given mutation set and illustrate them visually with red, yellow, and green in a *heatmap*.

For a given set of mutations $M = \{m_1, m_2, \ldots, m_n\}$. , $n \geq 1$, we would like to explore and visualize the set of functions that are highly important for the given mutations. Firstly, we retrieve all the genes associated with each mutation in the given mutation set. Then, we extract all the *bp* functions from the *pb* aspect (taxonomy) of the *GO* for all the genes associated with the mutations. That is, for each gene g_i associated with some mutation $M = \{m_1, m_2,$ in the set M., we obtain the *pb* annotations from the gene ontology annotation database (*GOA_human* [14]). We illustrate the important set of functions (*bp* function terms) using *heatmap* visualization. The most important functions will be in (dark) red.

3 Data, Experiments, and Result

The main goals and contributions of this work are: *i-* to identify the most represented and most important *biological process* functions of disease mutations; and ii- to visualize them proficiently and effectively with *heatmaps* to assist and simplify functional analysis of mutations. For the first goal, we used two of the most commonly used mutation databases.

3.1 Gene and Mutation Data

We utilize the following sources for mutation data and gene information in this work:

– We used the following two sources for mutation data: UniProt *humsavar* mutation data; and *Clinvar*: which is part of NCBI [1–3].
– For annotations, we used the *Gene Ontology* (GO), which is widely used as the main source of functional annotations of all genetic data [12, 14–16, 21].
– For human gene functional annotations, we relied on the *GOA_human* database [14] {*note: GOA_human contains ~ 600K annotations*}.

The gene functional annotations from the gene ontology (GOA) for all human genes (*GOA_Human*), using only the *bp* taxonomy, contains:

Goa_human total # of bp annotations	152396
Goa_human total # of bp annotated genes	17716

3.2 Result with *UniProt Humsavar* Mutations Data

The *Uniprot HumSavar* data contains ~78K mutations as follows:

Disease mutations: 30485
Polymorphisms: 40032
Unclassified variants: 7934
Total: 78,451

We used 2450 genes from the *humsavar* disease mutations database.

For the first experiment, we analyzed the top 100 genes in *humsavar* that are highly represented in the disease mutations compared to the polymorphisms. Then, we extracted all the *bp* functions of these genes. The genes and *bp* functions are illustrated in Fig. 1(a) and Fig. 1(b) respectively. Figure 1(b) illustrates the *heatmap* visualization of the top 20 bp functions (from the *GO*) associated with the highest represented genes enriched in the disease/pathogenic mutations in the *humsavar* database.

For the highest 50 represented genes in *humsavar*, we found 730 bp functions, with a total of 1,171 gene-bp pairs. The top 50 bp functions associated with these 50 genes are illustrated in Fig. 1(c). Then, we analyzed the top 100 represented genes: we found 1,364 bp functions, with a total of 2,396 gene-*bp* pairs. To identify the most common *bp* functions associated with these 100 genes, we illustrated them in Fig. 1(d). Then, we compiled and sorted all disease mutations based on diseases, for a total of 4,247 diseases and 30,182 mutations (we found 303 mutations not assigned to specific diseases) with the following details:

Top 10 diseases have 2253 mutations
Top 20 diseases 3513 mutations
Top 50 diseases 6398 mutations
Top 100 diseases 9625 mutations

By analyzing the top 100 diseases associated with 9625 mutations, we found 119 genes, which include 1,677 bp functions (for these 119 genes), and 3,104 bp-gene pairs. The results are illustrated in Fig. 1(e). As shown in Fig. 1(e), the heatmap is a convenient way to visualize and explore the most important and most common *pb* functions associated with these highly *mutation-populated* diseases.

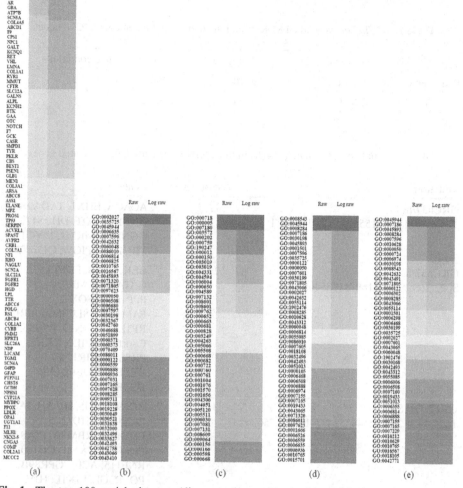

Fig. 1. The top 100 enriched genes (disease-polymorphism), along with the top bp functions associated with the genes in the *Humsavar* dataset

3.3 Results with the *ClinVar* Mutation Database

The *ClinVar* database contains more than 82,000 mutations divided into two classes, *benign* and *pathogenic* as shown Table 1. We conducted experiments using the *ClinVar* mutations to analyze and explore the most important and most represented *bp* functions among the mutations. We are mainly interested in disease mutations. We used the benign mutations as a control group for the analysis with the disease mutations to identify the highly represented *bp* functions. Table 2 shows a sample of three mutations from *ClinVar*. These are pathogenic mutations associated with multiple genes as shown in the table.

Table 1. In *ClinVar* we studied 82 K mutations that are pathogenic with 9270 genes

Type	# of mutations	# of genes	# of associated conditions
Pathogenic	82321	9270	7473
Benign	72070	9044	2572

Table 2. Sample of three mutations (from the *ClinVar* db) with their associated genes

Mutation	# of genes	Genes
GRCh37/hg19 17q12(chr17:34815551–36307189) × 1	12	ACACA, LHX1, TADA2A, HNF1B, AATF, DHRS11, GGNBP2, MRM1, MYO19, TBC1D3F, PIGW, C17orf78
GRCh37/hg19 8p23.1(chr8:8093169–11935465) × 1	21	BLK, CTSB, FDFT1, GATA4, MTMR9, PPP1R3B, SOX7, FAM167A, SLC35G5, ERI1, RP1L1, CLDN23, PRSS55, C8orf74, NEIL2, XKR6, PRSS51, MIR124-1, DEFB135, DEFB136, DEFB134
GRCh37/hg19 1q21.1–21.2(chr1:144368497–148636756) × 1	31	BCL9, FMO5, GJA5, GJA8, PEX11B, CHD1L, PDE4DIP, RBM8A, PIAS3, POLR3C, TXNIP, CD160, RNF115, ACP6, GPR89B, POLR3GL, LIX1L, HJV, ANKRD35, NBPF12, NBPF11, NUDT17, NBPF15, ANKRD34A, TCH2NLA, NBPF9, PPIAL4D, GPR89A, NBPF10, TRN-GTT9-1, TRQ-CTG3-1

We analyzed the top 10 mutations having the highest number of genes in *ClinVar* (see Table 2). These 10 mutations are associated with a large number of *bp* functions ranging from 2,810 bp functions to 4,607 bp functions per mutation. The top *bp* functions for the first ten mutations are illustrated in Figure 2.

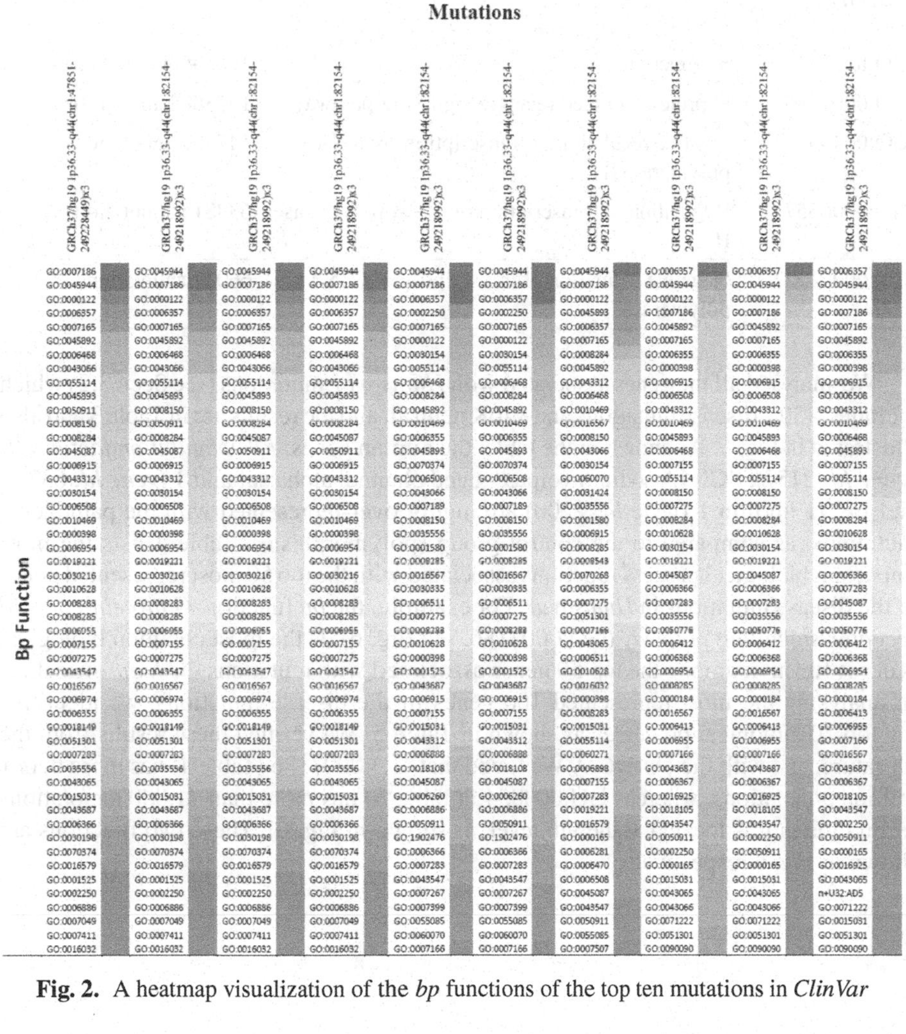

Fig. 2. A heatmap visualization of the *bp* functions of the top ten mutations in *ClinVar*

4 Results and Discussion

Figure 3 and Fig. 4 illustrate the most represented *bp* functions among the top *Clin-Var* disease mutations associated with the greatest number of genes. We identified the following *bp* functions are highly represented among the top 10 mutations (Figs. 3 and 4):

GO Id	*bp* function	Total # of annotations
GO:0007186	G protein-coupled receptor signaling pathway	1073888 annotations
GO:0045944	Positive regulation of transcription by RNA polymerase II	147360 annotations
GO:0006357	Regulation of transcription by RNA polymerase II	639814 annotations
GO:0000122	Negative regulation of transcription by RNA polymerase II	66037 annotations

We analysed all the genes associated with *Humsavar* mutations (see Sect. 3.2) which includes ~ 30K disease mutations and ~40K neutral, and the results are in Table 3 and also illustrated in Fig. 5. As shown in the table, the human genes: *F8 coagulation factor VIII* (*gene Id:* 2157), SCN1A sodium voltage-gated channel alpha subunit 1 (*gene Id*: 6323), and FBN1 fibrillin 1 (*gene Id*: 2200) are highly over-represented with the pathogenic mutations, as compared to the control group (polymorphism). Table 4 lists the most important biological process functions associated with the top 50 most represented genes in the disease mutations in *Humsavar*. For example, the *bp* function, *G protein-coupled receptor signalling pathway* (GO:0007186, see Fig. 6), is the most common biological process function among the top 50 genes associated with mutations the *humsavar* data. In analysing the most represented 100 genes in the disease mutations, we identified the top 20 biological processes as shown in Table 5. We verified these results from the literature, and with the *QuickGO*. We found that GO:0007186 (Table 4) has more than 1 million annotations, whereas GO:0008543 (Table 5) has more than 11,000 annotations (even though it is located at the 9th level, as shown in Fig. 6). These two functions are shown in Fig. 6 respectively.

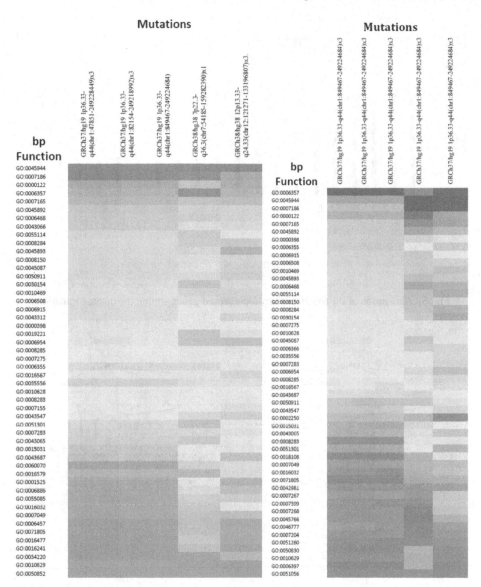

Fig. 3. The most highly represented *bp* functions of the top 5 *ClinVar* disease mutations with the highest number of genes

Fig. 4. The most highly represented *bp* functions of the second five disease mutations in *ClinVar*

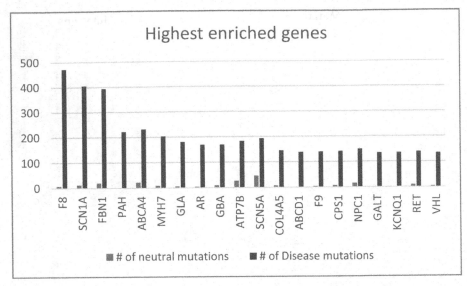

Fig. 5. Illustration of the top 20 highest represented genes among the disease mutations

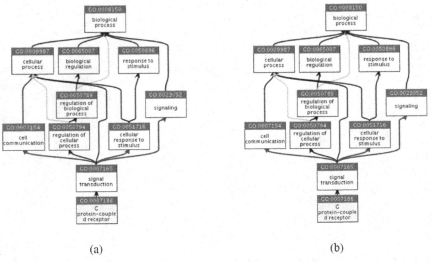

(a)   (b)

Fig. 6. Illustration of two of the most important biological process (*bp*) functions from the gene ontology (*bp taxonomy*) from *QuckGO*

Table 3. The top 20 genes that are highly represented in the disease mutations compared with polymorphism in *Humsavar*

Gene name	# of neutral mutations	# of Disease mutations	Diff: Disease - Polymorphism
F8	7	472	465
SCN1A	12	405	393
FBN1	19	393	374
PAH	1	222	221
ABCA4	22	231	209
MYH7	9	203	194
GLA	6	180	174
AR	4	169	165
GBA	9	169	160
ATP7B	26	183	157
SCN5A	46	193	147
COL4A5	7	145	138
ABCD1	1	138	137
F9	4	139	135
CPS1	7	141	134
NPC1	16	150	134
GALT	1	134	133
KCNQ1	2	135	133
RET	9	139	130
VHL	4	134	130

Table 4. The top 20 bp functions for the top 50 most represented genes in *Humsavar*

Bp function		Count
GO:0007186	G protein-coupled receptor signaling pathway	15
GO:0000050	Urea cycle	12
GO:0071805	Potassium ion transmembrane transport	12
GO:0035725	Sodium ion transmembrane transport	11
GO:0002027	Regulation of heart rate	10
GO:0007596	Blood coagulation	9

(*continued*)

Table 4. (*continued*)

Bp function		Count
GO:1902476	Chloride transmembrane transport	9
GO:0000122	Negative regulation of transcription by RNA polymerase II	8
GO:0001501	Skeletal system development	7
GO:0030198	Extracellular matrix organization	7
GO:0030199	Collagen fibril organization	7
GO:0043312	Neutrophil degranulation	7
GO:0045944	Positive regulation of transcription by RNA polymerase II	7
GO:0060048	Cardiac muscle contraction	7
GO:0006508	Proteolysis	6
GO:0045893	Positive regulation of transcription DNA-templated	6
GO:0071320	Cellular response to cAMP	6
GO:0086010	Membrane depolarization during action potential	6
GO:0086011	Membrane repolarization during action potential	6
GO:0097623	Potassium ion export across plasma membrane	6

Table 5. The top 20 bp functions for the top 100 most represented genes in *Humsavar*

Bp function		Count
GO:0008543	Fibroblast growth factor receptor signaling pathway	32
GO:0045944	Positive regulation of transcription by RNA polymerase II	28
GO:0008284	Positive regulation of cell population proliferation	25
GO:0007186	G protein-coupled receptor signaling pathway	19
GO:0030198	Extracellular matrix organization	16
GO:0045893	Positive regulation of transcription, DNA-templated	16
GO:0,001,501	Skeletal system development	14
GO:0007596	Blood coagulation	14
GO:0035725	Sodium ion transmembrane transport	14
GO:0000122	Negative regulation of transcription by RNA polymerase II	13
GO:0000050	Urea cycle	12
GO:0007601	Visual perception	12

(*continued*)

Table 5. (*continued*)

Bp function		Count
GO:0030199	Collagen fibril organization	12
GO:0071805	Potassium ion transmembrane transport	12
GO:0043066	Negative regulation of apoptotic process	11
GO:0002027	Regulation of heart rate	10
GO:0042632	Cholesterol homeostasis	10
GO:0055114	Oxidation-reduction process	10
GO:1902476	Chloride transmembrane transport	10
GO:0008285	Negative regulation of cell population proliferation	9

5 Conclusions

We presented the results of our work in finding the most important biological process functions that can be associated with disease mutations. We analysed the most commonly occurring biological process functions with disease mutations, as compared with polymorphisms or *benign* mutations. We found that highly represented functions are very informative and can be explored effectively with heatmap visualization. Identifying the most important functions for a given disease mutation, or a set of mutations, can lead to advances in other areas like gene-disease association analysis and mutation pathogenicity analysis. We found that identifying some highly represented and enriched genetic functions for disease mutations to be very helpful and informative for other important tasks like drug discovery and genetic discovery for disease treatment. This shall ultimately lead to advances in medical applications related to genetic treatment of hard disease like certain types of cancers.

References

1. Stephens, Z., Wang, C., Iyer, R., Kocher, J.: Detection and visualization of complex structural variants from long reads. BMC Bioinform. **19**(Suppl 20) (2018). https://doi.org/10.1186/s12 859-018-2539-x
2. Landrum, M.J., Lee, J.M., Benson, M., Brown, G., Chao, C., Chitipiralla, S., et. al.: ClinVar: public archive of interpretations of clinically relevant variants. Nucleic Acids Res. **44**(Database issue), D862–D868 (2016). https://doi.org/10.1093/nar/gkv1222
3. Stenson, P.D., Ball, E.V., Mort, M., Phillips, A.D., Shiel, J.A., Thomas, N.S.T., Abeysinghe, S., Krawczak, M., Cooper, D.N.: Human Gene Mutation Database (HGMD): 2003 update. Hum. Mutat. **21**, 577–581 (2003)
4. Bailey, M.H., Tokheim, C., Porta-Pardo, E., et al.: Comprehensive characterization of cancer driver genes and mutations. Cell **173**, 371–385 (2018)
5. Krawczak, M., Cooper, N.D.: The human gene mutation database. Trends Genet **13**, 121–122 (1997)
6. Genetic Home Reference GHR, US National Library of Medicine, NIH. Retrieved September 2018; https://ghr.nlm.nih.gov/

7. Genetic Alliance; The New York-Mid-Atlantic Consortium for Genetic and Newborn Screening Services. Understanding Genetics: A New York, Mid-Atlantic Guide for Patients and Health Professionals. Washington (DC): Genetic Alliance; 2009. Available from: https://www.ncbi.nlm.nih.gov/books/NBK115568/
8. GOTermFinder. https://go.princeton.edu/cgi-bin/GOTermFinder.
9. Kordopati, V., Salhi, A., Razali, R., Radovanovic, A., Tifratene, F., Uludag, M., Bajic, V.B.: DES-mutation: system for exploring links of mutations and diseases. Sci. Rep. **8**, 13359 (2018). http://doi.org/https://doi.org/10.1038/s41598-018-31439-w
10. Opap, K., Mulder, N.:. Recent advances in predicting gene–disease associations. F1000Research J. **6**, 578 (2017). https://doi.org/10.12688/f1000research.10788.1
11. Calabrese, R., Capriotti, E., Fariselli, P., Martelli, P.L., Casadio, R.: Functional annotations improve the predictive score of human disease-related mutations in proteins. Hum. Mutat. **30**(8), 1237–1244 (2009). https://doi.org/10.1002/humu.21047
12. Capriotti, E., Martelli, P.L., Fariselli, P., Casadio, R.: Blind prediction of deleterious amino acid variations with SNPs&GO. Hum. Mutat. **38**(9), 1064–1071 (2017). https://doi.org/10.1002/humu.23179
13. Kreft, L., Turan, D., Hulstaert‡§, N., Botzki†, A., Martens, L., Vandermarliere, E.: Scop3D: online visualization of mutation rates on protein structure. J. Proteome Res. **18**(2), 765–769 (2019). https://doi.org/10.1021/acs.jproteome.8b00681
14. Gene Ontology Annotation (GOA) Database: https://www.ebi.ac.uk/GOA
15. Araujo, F.A., Barh, D., Silva, A., Guimaraes, L., Juca Ramos, RT.: GO FEAT: a rapid web-based functional annotation tool for genomic and transcriptomic data. Sci. Rep. **8**, 1794 (2018)
16. Al-Mubaid, H.: Gene mutation analysis for functional annotations using graph heuristics. In: Proceedings of IEEE CIBCB (2019)
17. Wang, M., Wei, L.: iFish: predicting the pathogenicity of human nonsynonymous variants using gene-specific/family-specific attributes and classifiers. Sci. Rep. **6**, 31321 (2016). https://doi.org/10.1038/srep31321
18. Online Mendelian Inheritance in Man, OMIM. McKusick-Nathans Institute of Genetic Medicine, Johns Hopkins University (Baltimore, MD) (2018). https://www.omim.org/
19. Butkiewicz, M., Blue, E., Leung, Y., Jian, X., Marcora, E., et.al.: Functional annotation of genomic variants in studies of late-onset Alzheimer's disease. Bioinformatics **34**(16), 2724–2731 (2018). https://doi.org/10.1093/bioinformatics/bty177
20. Liu, X., Wu, C., Li, C., Boerwinkle, E.: dbNSFP v3.0: a one-stop database of functional predictions and annotations for human nonsynonymous and splice-site SNVs. Hum. Mutat. **37**(3), 235–241 (2016)
21. Al-Mubaid, H.: Analysis of gene variants for functional annotations. In: Proceedings of IEEE CIBCB (2019)
22. Xianfeng, L., Leisheng, S., Yan, W., Jianing, Z., Xiaolu, Z., Huajing, T., Xiaohui, S., Haonan, Y., Shasha, R., MingKun, L.: OncoBase: a platform for decoding regulatory somatic mutations in human cancers. Nucleic Acids Res. **47**(D1), D1044–D1055 (2019). https://doi.org/10.1093/nar/gky1139
23. Wang, K., Li, M., Hakonarson, H.: ANNOVAR: functional annotation of genetic variants from high-throughput sequencing data. Nucleic Acids Res. 38, e164 (2010)
24. Zhong, L.X., Kun, S., Jing, Q., Jing, C., Denise, Y.: Non-syndromic hearing loss and high-throughput strategies to decipher its genetic heterogeneity. J. Otol. **8**(1), 6–14 (2013)
25. Ng, P.K.S., Li, J., Jeong, K.J., Shao, S., et al.: Systematic functional annotation of somatic mutations in cancer. Cancer Cell. **33**(3), 450–462.e10 (2018). https://doi.org/10.1016/j.ccell.2018.01.021
26. Ward, M.O., Grinstein, G., Keim, D.: Interactive Data Visualization Foundations, Techniques, and Applications, 2nd ed. Taylor and Francis (2015) https://doi.org/10.1201/b18379

27. Batch, A., Elmqvist, N.: The interactive visualization gap in initial exploratory data analysis. IEEE Trans. Vis. Comput. Graph. **24**(1), 278–287 (2018)
28. Zyla, J., Marczyk, M., Weiner, J., Polanska, J.: Ranking metrics in gene set enrichment analysis: do they matter? BMC Bioinform. (May 2017). https://doi.org/10.1186/s12859-017-1674-0
29. Fernandez, N.F., Gundersen, G.W., Rahman, A., Grimes, M.L., Rikova, K., Hornbeck, P., Ma'ayan, A.: Clustergrammer, a web-based heatmap visualization and analysis tool for high-dimensional biological data. Sci. Data. **4**, 1–12 (2017). https://doi.org/10.1038/sdata.2017.151
30. López-Urrutia, E., Salazar-Rojas, V., Brito-Elías, L., et al.: BRCA mutations: is everything said? Breast Cancer Res Treat (2018). https://doi.org/10.1007/s10549-018-4986-5
31. Doughty, E., Kertesz-Farkas, A., Bodenreider, O., Thompson, G., Adadey, A., Peterson, T., Kann, M.G.: Toward an automatic method for extracting cancer- and other disease-related point mutations from the biomedical literature. Bioinformatics **27**(3), 408–415 (2011). http://doi.org/https://doi.org/10.1093/bioinformatics/btq667
32. Karpatne, A., et al.: Theory-guided data science: a new paradigm for scientific discovery from data. IEEE Trans. Knowl. Data Eng. **29**(10), 2318–2331 (2017)
33. Sarker, I.H., et al.: Cybersecurity data science: an overview from machine learning perspective. J. Big Data **7**, 41 (2020). https://doi.org/10.1186/s40537-020-00318-5
34. Cao, L.: Data science: challenges and directions. Commun. ACM **60**(8), 59–68 (2017)
35. Feng, J., Xu, H., Mannor, S., Yan, S.: Robust logistic regression and classification. Adv. Neural Inform. Proc. Syst. **27** (NIPS 2014)
36. Rizk, A., Elragal, A.: Data science: developing theoretical contributions in information systems via text analytics. J. Big Data **7**, 7 (2020). https://doi.org/10.1186/s40537-019-0280-6

Applied Data Science, Artificial Intelligence, and Data Engineering

Characterizing In-Situ Solar Wind Observations Using Clustering Methods

D. Carpenter[1]([✉]) [ID], L. Zhao[1] [ID], S. T. Lepri[1] [ID], and H. Han[2] [ID]

[1] Department of Climate and Space Sciences and Engineering,
University of Michigan, Ann Arbor, MI 48109, USA
dcar@umich.edu

[2] Department of Computer Science, Rogers School of Engineering and Computer Science, Baylor University, Waco, TX 76798, USA

Abstract. The Sun's atmosphere is a hot, high pressure, partially ionized gas, that overcomes the Sun's gravity to create a flow of plasma that expands out into the solar system: a phenomena that is called the solar wind. The solar wind comprises the continuous and dynamic streams of plasma released from the Sun. Some of the physical properties of these plasma streams (composition, charge state, proton entropy) can be determined in the solar corona and remain non-evolving as the solar wind parcels expand into the heliosphere. These properties can be used to connect in-situ measurements to their coronal origins. Determining how solar wind parcels differ based their coronal origins will reveal the nature of physical processes (such as heating and acceleration) involved in their formation. Studies up to this point have largely relied upon statistical methods, characterizing the wind into groups such as fast and slow wind. Other methods have been used to detect signatures that represent transient events, such as interplanetary coronal mass ejections (ICMEs). The boundaries representing physical distinctions between the groups usually have included an aspect of them that was subjective. In the past few years, there has been a growing push to use machine learning in the field of heliophysics, in the form of (including but not limited to) predicting conditions in the solar wind/time series regression, identifying coronal features/events in solar images, and to serve as a tool to reduce subjective bias when determining solar wind groups. Here, we examine how machine learning, applied to several case studies of the solar wind, has the potential to link the physical properties measured in-situ to the origin of the wind at the Sun. We evaluate the robustness of the results of two different methods, comparing their performance. We discuss caveats that may arise when applying such techniques. Finally, we identify their strengths and define what aspects would be beneficial in continuing to develop machine learning approaches to this field.

Keywords: Solar wind · Classification · Machine Learning · Heliophysics · Interplanetary physics · Interplanetary turbulence · Interplanetary magnetic fields · Solar coronal mass ejections · In-situ data analysis

Supported by University of Michigan Department of Climate and Space Sciences and Engineering.

1   Introduction: The Solar Wind

Space plasma is a substance which consists of a partially or fully ionized gas, created by the high temperature environment of a star. The solar plasma consists of protons (ionized hydrogen), helium, and trace heavier elements. Solar wind particles have varying degrees of ionization (here, their *charge state*). The abundance elements or individual charge states are typically represented as a ratio of densities (*elemental abundance ratios* and *charge state ratios*, respectively). Some widely used examples of these are with oxygen: O^{7+}/O^{6+} and carbon: C^{6+}/C^{5+} (the plus symbol indicates the number of positive charges in the atom) [8]. These solar ions are heated, accelerated, and released from the Sun in the form of a constant, dynamic stream known as the *solar wind*. The solar wind expands nearly radially from the Sun as it overcomes its gravitational pull, and flows out into the solar system to form the plasma bubble that is known as the *heliosphere*. Periodically, disturbances on the Sun and in the solar wind, such as interplanetary coronal mass ejections (hereafter: ICMEs), solar flares, and high speed streams will propagate outward into the heliosphere, and these disturbances can encounter and interact with planetary magnetospheres and atmospheres. These ever-changing conditions influenced by the Sun are referred to as *space weather phenomena* [14].

Space weather's impact at Earth is a part of the fundamental processes that operate in its magnetosphere. Energy, stored in the magnetotail, is released due to the build up through interactions with space weather disturbances, and is transmitted along the magnetic field lines, funnelling at Earth's magnetic poles [1]. Energized particles enter the upper atmosphere here, and they excite atmospheric gases which manifests as a glow: the aurorae. Space weather events can lead to major disruptions of Earth's magnetosphere, which are referred to as *geomagnetic storms* [1].

Geomagnetic storms pose a continual and significant risk to the technological presence in the magnetosphere. The primary risk for technological systems is to their onboard electronics and power systems [1]. Disturbances in the local space environment can also affect spacecraft communications and ranging, and can enhance drag on satellites. This can lead to premature de-orbiting and loss of spacecraft [1,7,9].

Space Weather also can have a consequential impact to life on the surface. FEMA's 2019 Federal Operating Concept for Impending Space Weather Events lists Space Weather as one of the top potentially catastrophic risks, explaining that a solar storm can cause "systemic cascading impacts, destroying infrastructure critical to the national economy and security" [4]. One of the most famous events is the long retold 1859 "Carrington Event". Twice as large as any solar storm in the past 500 years, it would lead to induced electrical currents in metal water pipes, sparks leaping from telegraph operating stations, and an auroral oval seen as low as Honolulu at (roughly) 21° N latitude [14]. It is predicted that if a Carrington-class solar storm were to hit Earth today, it could cost between 1–2 trillion USD to the U.S. in damages (2008 estimate) [6]. Even today, space weather greatly impacts high latitude nations in both atmospheric and surface effects–to the point that such considerations are integrated into their infrastructure funding [1,4,7]. It is without question that our reliance, utilization, and exploration of the space environment will require a greater understanding of real-time and predicted solar wind conditions. This work will focus on the former, although the latter represents one of the leading applications of machine learning methods in heliophysics. Here, we will focus on solar wind characterization as a means of better understanding connections to the solar corona, and to qualify in-situ conditions in a set of "states" that allow for improved assessment of geomagnetic impact.

Solar wind measurements are largely grouped into two categories: whether the bulk proton speed is relatively slow or fast [2,12,17,18]. This allows for an effective, simple separation of wind types [5]. Unfortunately, the two-group solar wind classification scheme is not necessarily unique, and ultimately up to subjective choice; there are considerable statistical differences between different types of solar wind when other physical parameters are included (e.g. x-ray emissions from coronal holes, elemental composition, and ion charge states) [17,18]. Aside from a wind speed categorization being less adequate for distinguishing solar source regions, it also does not account for co-rotating interaction regions (CIRs)–"compression regions when high speed streams meet slower speed streams". These CIRs, particularly in the ecliptic, can cause slow wind to speed up and faster wind to slow down, muddying possible signatures to coronal sources [2]. Because of the statistical complexity of solar wind data, and the subjectivity introduced through standard procedures, machine learning has been implemented as a tool to help solve the problem as to how in-situ solar wind measurements could be objectively characterized.

2   K-Means: Spatially Separate Clusters

When introducing machine learning to this problem, it is important to note that there are many different kinds of clustering methods that have been developed–one such method that limits subjective choice is using the K-Means algorithm. Because there is no label placed a-priori upon the training data, the algorithm is considered *unsupervised* (supervised methods such as neural networks often have "flags" in the training data that help them determine a weighting function. The hyperparameter that is selected by the user in this algorithm is the value 'K', which is the number of clusters that will be in the output of K-Means. From this, there will be 'K' cluster centers, referred to as centroids, that represent all of the nearest points. The K-Means algorithm originates from the 1957 work by Hugo Steinhaus [13]. A pseudocode can be found in Algorithm 1. K-Means starts by randomly[1] assigning a "k" amount of cluster centroids onto the high dimensional data. Then, the distance from each point to the centroid is calculated. The clusters are formed by labeling data samples with the centroid has the *least distance* to that sample (e.g. each cluster has points that are "closest" to the

---

[1] This initialization can be made with the partial help of an algorithm, in order to improve the quality of the results.

each centroid). At this point, there are k-number of groups of data, each with a label corresponding to the cluster which that data belongs. Now, the average value of each dimension in each cluster is taken, and that average becomes the *new* centroid position for that cluster. With the centroid reassigned, the closest points are taken once again (possibly being a list somewhat different from before), and the process repeats–until the centroid positions cease "shifting" very much (or after a certain number of repititions). At this point, the clusters are considered "stable", and the resulting clusters are the final output of the k-means algorithm. Strengths of this algorithm relate to how the boundaries between characterized groups are determined autonomously by the data (e.g. the method is *data driven*). The approach is also very simple in terms of how machine learning algorithms operate: it is both transparent and easily re-created.

Algorithm 1: K-Means (X, k)

1 **Input:**
2     $X = [x_1, x_2, \ldots x_N]$: input data.
3   k: the number of clusters in which to organize the data.

4 **Output:**
5     $C = \{c_1, c_2, \ldots, c_k\}$: cluster centers.

6 **Begin.**

7   Randomly select k data points as initial cluster centers.

8 **Repeat.**
9   Reinitialize all partition subsets as empty.
10    $S_1 = S_2 = \ldots = S_k = \{\}$.
11   Assign each data to the closest cluster center.
12   **For** $i \in \{1, 2, \ldots. N\}$ **do:**

13        $l = \text{argmin}_{\{j \in 1,2,\ldots,k\}} \|\mathbf{x}_i - c_j\|^2$;
14        $S_l = S_l \cup \{x_i\}$;
15   **End**

16   Update cluster centers.
17   **For** $j \in \{1, 2, \ldots, k\}$ **do:**
18        $c_j = \sum_{i \in \{1,2,\ldots,N\} x_i \in S_j} x_j / |S_j|$;
19   **End**
20   **Until** *the cluster assignment converges;*

21 **End**

In the context of heliophysics, this algorithm was applied in a study by Roberts et al. (2020) [12]. The goal of this paper was to explore how an objective classification scheme could identify different states of solar wind and explore how those ML identified solar wind states compared to the more subjectively classified solar wind types that have been the typical standard in heliophysics. Aside from the measured physical parameters, non-local variables to quantify solar wind turbulence such as cross helicity and residual energy were introduced. The cross helicity is a "measure of how purely the mode of the wind fluctuations is that of an Alfvén wave ($\sigma_C = \pm 1$)". Such waves tend to propagate outward from the Sun, and the sign is determined by the direction of the mean magnetic field [12]. They initially started with a K = 8 category scheme (including the cross helicity and residual energy), and then a K = 4 scheme, based on the Xu & Borovsky 2015 variables (to see how their groups would compare to those found by this method). In this case, Euclidean distance metric was used as a starting point for this data. Ultimately, they were able to attribute solar source features to the output of their K-Means, and even were able to isolate a group which they believed aligned closely with ICMEs. This work shows the utility of K-Means in classifying solar wind states and the potential for future application as an automated "scientist in the loop" when more traditional identification is not feasible (or when to determine when is best for a satellite to use a certain mode of operation) [12]. In our work, we will examine the stability and consistency this approach, and consider the best practice when the applications of K-Means is concerned.

K-means serves as a great starting point because of its ease of use and its suitability for solar wind data behavior [5]. Additionally, it is a fairly intuitive unsupervised learning algorithm, making it somewhat feasible to trace and diagnose. The algorithm is run using the magnetic field strength, plasma density, proton speed, O^{7+}/O^{6+} charge state ratio, the relative abundance of iron (Fe/O), the average charge state of iron, as well as the time correlations known as cross helicity and residual energy, with measurements taken by the Advanced Composition Explorer (ACE) in a time span that featured a high amount of ICME's, dating from November of 2002 to May of 2004. The data was pre-processed (normalizing from 0 to 1, associated with the minimum and maximum of each physical parameter, respectively), and the non-local variables cross helicity and residual energy were calculated to form a time series. There are several options for choice of distance metric, so Euclidean was selected as (quite literally) the most straightforward; although there may be merit to using k-medians, according to a few atmospheric in-situ particle studies. Because this is left to the user's choice, it is considered subjective, although the choice of a straight-line distance between points is standard practice when exploratory work like this is being done.

Once this is done, the delineation of groups shouldn't be along any single parameter (which would imply that all the other parameter spaces were uniform). This can be used to assess the quality of the clusters. To verify the validity of this demonstration, the results were compared and visualized in a subset of the data ranging from May to September of 2003. The differences in trials will serve to give a rough evaluation at how strong the groups identified by K-Means are (their stability), and demonstrate how well K-Means can classify solar wind.

Starting multiple trials of K-Means describes a larger story about the considerations that should be taken when using techniques such as this, shown in Fig. 1. It is seen that individually, they show some distinctions between the solar wind types (labelled by numeric value and colored to distinguish types). We see overlap, which is important, because it shows for instance that the training using the nonlocal variables didn't control the behavior of the data-if we only saw clean, straight-line boundaries in these projections, it would imply that the other parameters were not affecting the result; however, over various trials with the same starting inputs, the resulting outputs differ quite a lot. The coloration is done to approximate the general "type" of cluster that it is associated with, as the numeric labels are un-connected to the characteristics of the groups (although, the cyan colored cluster is used to highlight an overlap that previously didn't exist (Fig. 1)). What this demonstrates is that the positions of the central values in this space, and the resulting qualities of the clusters can have inconsistencies-the algorithm on this data doesn't behave the same way every time.

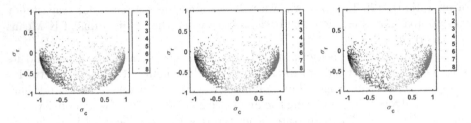

Fig. 1. Three trials of K-Means (k = 8) on ACE data. The data is projected onto two of the training variables: cross helicity σ_C and residual energy σ_R

For K-Means, the clusters display similarities (in general) to the ones listed before (and the variations over the trials still persist). As a result, groups of data which accounts for different characteristics of solar wind can be formed. When projected onto physical variables, these groups can show distinctions from one another and–when examining properties of the centroid positions–exhibit "typical" characteristics of the group they have been clustered into. The method is simple, and quick to run which makes subsequent trials easily attained. There are a couple trends to note when using k-means to perform cluster analysis to the goal of understanding solar wind origins. The first is that the coronal feature in which a group belongs is dependent on the user's interpretation of parameter values. As discussed later, there are a few ways this can be analyzed, like in the case of ICME's. They can also be compared to classically attained solar wind groups to search for how well a data driven feature compares. Future work would require a potential statistical analysis to understand how much of the data controls each group, and why a certain k value should be chosen to seek said groups; Heidrich-Meisner & Wimmer-Schweingruber (2018) demonstrated by justifying their choice heuristically. The choice of k value remains subjective, but there could be further steps explaining why we constrain to a certain number of groups. Lastly, there is the fact that these clusters appear to be sensitive to the initial centroid starting positions, which are random. As multiple trials are run, the boundaries of the groups appear to shift, indicating that the stability of these clusters is not as strong as previously considered. This is possibly due to the behavior of solar wind data making classification in this exact parameter space difficult, or suggests the notion that the k-value may be too high (Xu & Borovsky parameters showing much more stability than the selected raw ACE set).

2.1 ICME Characterization

There is a unique ability demonstrated when applying machine learning techniques to capture behavior of ICME activity, in a way that is registered in the time series analysis. Here, the potential for how ICME's could be captured using unsupervised techniques is demonstrated, and new ways solar wind observations could be characterized (and how ICME's would potentially perform with that metric) are proposed.

The first step was to revisit k-means clustering. Using expected temperature, Alfven Speed, and Proton Entropy (taking the log base 10 of these values), the Xu and Borovsky variables [17] are applied, in the same date ranges as before. Information regarding ICMEs came from the 2010 Richardson and Cane study which catalogues and summarizes the denoted average values [11]. Instead of 8, 4 clusters were used when originally making separations in clusters in a three-dimensional parameter space.

Using the centroid positions, the result is projected onto the wind speed and O^{7+}/O^{6+} ratio time series (Fig. 2). Of note, there was a remarkable isolation of ICME periods evidenced in the proton speed time series and the charge state ratio of oxygen. In Fig. 2, normalized components of the magnetic field indicate magnetic polarity, and in the O^{7+}/O^{6+} charge state ratio time series, the blue horizontal line is a threshold that indicates enhanced heavy ion phases from a 2009 paper by Zhao et al. [19]. There are complementary interpretations of these results afforded by how strongly these clusters have separated; for example, the violet group tends to have high wind speeds and very low O^{7+}/O^{6+}, making it a good candidate for the coronal hole wind. The lime green cluster is the proposed ICME group, and what is seen in Fig. 2 is a general trend where each published Richardson ICME is in close proximity to this cluster. This corresponds well with the heavy ion ratio plot, where large peaks are also greatly aligned to the listed ICME events. This isn't perfect, and there are instances where the lime green group is absent of the ICME series, but this overlap was determined in a hands-off approach (nothing told k-means to look for ICME events, but rather the training set was supplied with a time frame of a relatively large amount of ICME's). In all, this motivates future work into closely examining how ICME groups may be extracted, and testing different scenarios that might improve our ability to identify and characterize ICME's.

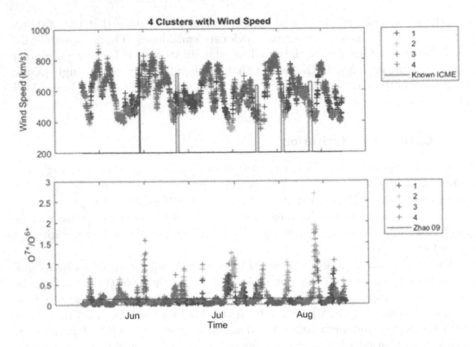

Fig. 2. Clusters from K-Means (K = 4) trained off Xu & Borovsky 2015 variables [17], projected onto solar wind speed (top) and O^{7+}/O^{6+} charge state ratio (bottom). Blue bars on the top series are averages and durations from the Richardson and Cane ICME list [11]. (Color figure online)

3　UMAP: Latent Features

In Uniform Manifold and Projection (UMAP), like K-Means, the physical parameters behave like spatial dimensions [2]. The practice of forming a lower-dimensional manifold on the higher dimensional dataset is the design philosophy behind the UMAP algorithm: to learn the topological structure of the manifold and to find a lower dimensional representation of the data that has "an equivalent topological structure". The result is the transformation of the input data into a "much simpler [representation] that still [captures] meaningful structural features of the original dataset" [2]. A manifold and projection scheme expresses that multidimensional dataset and its behavior in "a much smaller number of latent features" [2]. In this new representation of the data, called the *reduction*, the groups can be clustered using unsupervised learning techniques. The consequence of trying to do this on a high dimensional and non-uniform dataset is that the manifold isn't exact; rather, it is a "fuzzy" simplification that uses stochastic gradient descent to form an approximation that is close to the true topological structure. The concept for UMAP originally came from the formulation by McInnes, Healy, and Melville [10]. Once a reduction is created to represent the input set, the data points must be assigned according to which group they belong to; UMAP does not do this by itself. Rather, it organizes the data to a lower dimensionality to primarily represent the structure of the data. To attain the clusters in this reduction the Hierarchical Density Based Spatial Clustering for Applications with Noise (HDBSCAN) algorithm will more objectively determine the boundaries–and even the total number–of clusters.

The HDBSCAN algorithm finds "dense" clusters, and "constructs a hierarchical tree of contiguous regions of density". Once all of the points fall into this hierarchical tree, it is simplified to yield a "single flat clustering" that labels each point as a "cluster identity, or as noise" [2]. This noise group is what HDBSCAN determines to be un-classifiable, as it doesn't quite fit well into the cluster identities it uses to form the hierarchy for the other groups. In an ideal case, K-Means and HDBSCAN can produce similar results; however, HDBSCAN can form non-linear boundaries for more complex cluster shapes such as rings, and crescents [3]. HDBSCAN does not have the number of clusters specified a-priori–this means that the algorithm is agnostic to the number of groups (of solar wind, in this case) that exist in the data. This improves the ability of provide exploratory work without biasing the approach with the number of groups that could be "sought out". For all of these reasons, HDBSCAN is ideal for handling the various topologies that arise from the reduction output of UMAP.

These algorithms were used in unison by Bloch et al. (2020) in order to apply them to the problem of characterizing in-situ solar wind data. In this study, they used the advantages of the Ulysses mission, which orbited in an inclined plane that took it to $\pm 80°$ in heliolatitude, allowing direct sampling of polar coronal hole wind [16]. This is important, as most satellites lie close to the ecliptic, where they receive more slow and equatorial coronal hole wind–using the Ulysses mission (in particular when it crossed a relatively wide latitude range in a brief period of time) allowed for the sampling of a more diverse set of wind data in which to classify. In Bloch et al. (2020), they applied two different methods: Bayesian Gaussian Mixture (BGM) scheme, and then a combination-approach of the UMAP & HDBSCAN. Their results were impressive, as the process yielded a representation of the data in a feature-space, and uniquely derived solar wind types that were well aligned with coronal features. There was a particular focus on making each step as objective as possible, even when it came to deciding the number of groups themselves (where K-Means is determined by the user). They found significant overlap in the identified solar wind types with the more traditional solar wind speed separation scheme, but disparities revealed morphological differences between coronal hole wind structure observed at high latitude and in the ecliptic, particularly when identified via the UMAP scheme. In our work, we examine more closely this UMAP scheme, because it was presented as a response their BGM scheme's drawbacks (so that the distribution of parameters would not be assumed, and to further remove points of subjectivity) [2].

By using data from the Ulysses mission, a wide latitudinal range was represented, as well as the selection in solar wind parameters that are considered non-evolving past the corona (e.g. average iron charge state, iron to oxygen ratio, alpha to proton ratio, etc.) [2]). Here, we test the robustness to input config-uration of their results when applied to the Ulysses data, and we discuss the potential implications of the hyperparameter selection.

The relatively wide range of solar latitudes that Ulysses covered can be examined directly in periods known as the fast latitudinal scans. This ran from the approximate dates of 15/08/94–20/08/95, 01/11/2000–01/11/2001, and 01/02/07–01/02/08. Classification could then be made with parameters that are considered non-evolving as the solar wind traveled to the distance of the Ulysses orbit. Namely, these are the O^{7+}/O^{6+} ratio, the proton entropy, the C6+/C5+ ratio, the average iron charge state, the iron to oxygen ratio, and the alpha to proton ratio. The observations were once again min-max normalized from 0 to 1, and UMAP was applied to uncover the underlying structure of a dataset by reducing its dimensionality to a small number of latent features. In effect, the machine learning algorithm "learned" the topological structure of the man-ifold. In training the UMAP algorithm, the hyperparameters used are shown in Table 1. According to Bloch et al., they chose their neighbourhood value in order of have a "balance between hyper-localised structure and totally-global struc-ture". They chose their minimum distance in order "[give] the algorithm the most freedom to accurately represent the data in the lower-dimensional space". The spread was chosen to "ensure that there is enough distance between clusters for the clustering algorithm to appropriately capture the results" [2]. Here, we vary the distance metric to test the robustness of the results.

It is here that should be noted that the rigorous testing of this algorithm in this context is incomplete–in a more ideal implementation, a score would be prescribed to the approach and hyperparameters would be more varied to more exhaustively determine the configuration for the machine learning algorithm. For example, the Normalized Mutual Information (NMI) score, upon iterating over combinations of inputs (which can be done programatically by using gridsearch), can provide more solid justification for these choices. With that said, part of the sensitivity of this approach to these changes can be seen by adjusting just one of the hyperparameters; here, the distance metric.

Shown in Fig. 3a, the correlation distance metric was used to obtain clean, easily identifiable clusters. How this metric operates is by taking the linear cor-relation between subsequent samples. To see how this affected the clustering, the classes were projected onto physical data was examined to test the robust-ness of the topological behavior of the data (as with the multiple trials of the k-means scheme). As a reminder, this is being done to use the data to make a division in the distributions of physical solar wind values-nothing about the choice in hyper-parameters is informed by this; however, this means that the way the unsupervised learning methods make that cut needs to be examined. This is done here by changing the distance metric parameter and comparing both the reductions and projections of cluster indices onto the physical parameters. What follows is an example of visualizing the method's robustness to hyperparameter input.

Table 1. Hyperparameters used in training UMAP.

Hyperparameter	Value
Number of neighbors	40
Metric	'Correlation', 'euclidean'
Number of components	2
Minimum distance	0
Spread	0.5

As seen in Fig. 3, the method formulated by Bloch et al. [2] is able to separate solar wind observations of the Ulysses fast latitude scans into two main groups (or clusters) of solar wind, and an additional "unclassifiable" group mixed between the two–all without a physical model or constraint being placed on the input information. Rather, these represent what behavior can be attributed to the data set as a whole, and then identified automatically using a clustering algorithm. In the reduction of the data, there is a very significant difference depending on which distance metric hyper-parameter is selected (comparing Fig. 3a to Fig. 3b). The result from applying Euclidean distance (which was done before) does not lend itself to an easily identifiable 2 cluster scheme, which is evidenced by the unclassifiable cluster being skewed, in this case towards the "fast" wind speed group (Fig. 3e vs 3f). These shapes bring up a lot of questions regarding the consequences of different reductions, which can at least be in part answered by applying the clustering algorithm, and seeing how the points attributed to each cluster behave when projected onto physical values.

What we can see when projecting the indices onto the physical parameter space in the form of contours is quite a lot of similarity–the blue group (cluster 1 in the Fig. 3(a,c,e) and (b,d,f)), when examining its general wind speed and proton entropy aligns well with what is expected of coronal hole features, and is fairly consistent in the points that are classified as such, despite how much the reduction changes between distance metrics. The story for the other, orange cluster (cluster 2), is mostly similar. As a reminder, this can be lumped into the "slower" wind–the exact boundaries of which are contested. There is a large overlap between the two distance metrics, with the only difference being a higher density of points for low O^{7+}/O^{6+} values, where the proton entropy is close to 1 (Fig. 3c, d). The inclusion of a correlation distance metric in the formation of the manifold slightly, but directly, changes how slower wind gets characterized. It generally represents what is expected for streamer belt wind, but its exact identity is more tenuous. This point is vital in the philosophy of trying to create a purely objective data science (ideally), and configuring how we want our machine learning algorithm to make that cut. There is more work to be done to try to rule out these variances, and to try and reinforce strong, consistent boundaries between wind modes that are replicable and invariant.

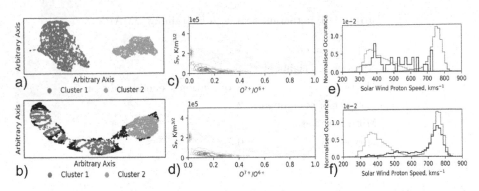

Fig. 3. UMAP and HDBSCAN on Ulysses Fast-Latitude Scan non-evolving parameters. The configurations are equivelant excepting the distance metric applied. The top row (a, c, e) uses correlation distance, and the bottom row uses Euclidean (b, d, f). The first column (a, b) are the reductions from UMAP, colored to the clusters detected by HDBSCAN. The middle column (c, d) is the projection of the clusters into contours on the training parameters O^{7+}/O^{6+} ratio, and proton entropy. The right column (e, f) are the normalized histogram distributions of the wind speed associated with each of the clusters

4 Conclusion

In this paper we examined the stability and consistency of using K-Means in the application of characterizing the solar wind, and we consider the best practice when the applications of K-Means is concerned. We also examined how UMAP and HBSCAN can identify solar wind features with limited subjectivity. We find that this type of feature selection process is a step in the right direction, but there is some room for improvement in terms of transparency and interpretation of results.

Machine learning algorithms still contain instances of subjectivity in the determination of solar wind states. In order to account for these, methodical reasoning must be given for each choice in hyperparameters-especially when interpreting the results. ICMEs can be captured via unsupervised learning methods, and they could potentially be formed as their own unique group when trying to classify solar wind. The machine learning methods covered here have unique strengths and weaknesses-all serving as a demonstration for the applications of machine learning to solar wind data. They serve as inspiration for developing novel techniques in which to improve upon; this has led to the schematic to apply in the future. This work, UMAP was used as a dimension reduction tool to aid clustering. UMAP is very stable (giving consistent results between trials), and is rather explainable (a very desirable aspect to machine learning algorithm, especially in the context of scientific applications. As shown, UMAP serves as a viable tool to approach this problem; however, UMAP is not the only type of algorithm that can return a reduction.

T-Distributed Stochastic Neighbor Embedding (t-SNE) is a means of accomplishing dimensionality reduction that has shown promise–matching even newer dimensional reduction algorithms which perform similar tasks when it comes to the *local behavior* of the groups we want to find. T-SNE is a variation of the previously established "Stochastic Neighbor Embedding" which "visualizes high-dimensional data by giving each datapoint a location in a two or three-dimensional map." This technique is "much easier to optimize [than Stochastic Neighbor Embedding], and produces significantly better visualizations by reducing the tendency to crowd points together in the center of the map. T-SNE is better than existing techniques at creating a single map that reveals structure at many different scales. This is particularly important for high-dimensional data that lie on several different, but related, low-dimensional manifolds, such as images of objects from multiple classes seen from multiple viewpoints" [15]. For our purposes, t-SNE will perform exceptionally, as we connect its output back to solar features.

This work has a wide scope in terms of methodologies and techniques applied to experimental analysis of solar wind properties, yet is focused in its goal to help answer how solar wind samples differ—what common properties can be ascertained by utilizing a larger range of input parameters. Solar wind has been classified into fast wind and slow wind most commonly by wind speed alone, but there is reason—especially in regards to slower wind—to explore what (in addition to the wind speed) also comprises the observations we see at 1 AU. Here, we have merely scratched the surface of what can be done, and the process can still be optimized and improved upon: machine learning shows promise as a tool in order to help answer outstanding questions in heliophysics such as what can be learned about the conditions where the wind originated from a parcel of solar wind plasma observed in the heliosphere. By relating groups of these observations to features in the corona (similarly to studies that had not used machine learning), a better understanding is gained about how one solar wind sample can differ from another.

References

1. Beetle, J.M.H., Rura, C.E., Simpson, D.G., Cohen, H.I., Moreas Filho, V.P., Uritsky, V.M.: A user's guide to the magnetically connected space weather system: a brief review. Front. Astron. Space Sci. **8**, 253 (2022) https://doi.org/10.3389/fspas.2021.786308
2. Bloch, T., Watt, C., Owens, M., McInnes, L., Macneil, A.R.: Data-driven classification of coronal hole and streamer belt solar wind. Sol. Phys. **295**(3), 1–29 (2020). https://doi.org/10.1007/s11207-020-01609-z
3. Campello, R.J.G.B., Moulavi, D., Sander, J.: Density-based clustering based on hierarchical density estimates. In: Pei, J., Tseng, V.S., Cao, L., Motoda, H., Xu, G. (eds.) PAKDD 2013. LNCS (LNAI), vol. 7819, pp. 160–172. Springer, Heidelberg (2013). https://doi.org/10.1007/978-3-642-37456-2_14
4. Federal Emergency Management Agency: Federal Operating Concept for Impending Space Weather Events (2019). https://www.fema.gov/sites/default/files/2020-07/fema_incident-annex_space-weather.pdf

5. Heidrich-Meisner, V., Wimmer-Schweingruber, R.: Solar wind classification via K-means clustering algorithm (2018)
6. Klein, C.: A perfect solar superstorm: the 1859 carrington event. https://www.history.com/news/a-perfect-solar-superstorm-the-1859-carrington-event. Accessed 20 June 2022
7. Knipp, D.J., Gannon, J.L.: The 2019 national space weather strategy and action plan and beyond. Space Weather **17**, 794–795 (2019)
8. Lepri, S.T., et al.: Solar wind heavy ions over solar cycle 23: ACE/SWICS measurements. ApJ **768**(1), 94 (2013)
9. Malik, T.: SpaceX says a geomagnetic storm just doomed 40 Starlink internet satellites. space.com, 8 February 2022. https://www.space.com/spacex-starlink-satellites-lost-geomagnetic-storm
10. McInnes, L., Healy, J., Melville, J.: UMAP: uniform manifold approximation and projection for dimension reduction. J. Open Sour. Softw. **3**(29), 861 (2018)
11. Richardson, I.G., Cane, H.V.: Near-earth interplanetary coronal mass ejections during solar cycle 23 (1996–2009). Catalog and summary of properties. Sol. Phys. **264**, 189–237 (2010)
12. Aaron Roberts, D., et al.: Objectively determining states of the solar wind using machine learning. ApJ **889**, 153 (2020)
13. Steinhaus, H.: Sur la division des corps matériels en parties. Bull. Acad. Polon. Sci. (in French) **4**(12), 801–804 (1957)
14. Temmer, M.: Space weather: the solar perspective. Living Rev. Sol. Phys. **18**, 4 (2021). https://doi-org.proxy.lib.umich.edu/10.1007/s41116-021-00030-3
15. van der Maaten, L., Hinton, G.: Visualizing data using t-SNE. J. Mach. Learn. Res. **9**, 2579–2605 (2008)
16. Wenzel, K.P., Marsden, R.G., Page, D.E., Smith, E.J.: The Ulysses mission. Astron. Astrophys. Suppl. **92**, 207 (1992)
17. Xu, F., Borovsky, J.E.: A new four-plasma categorization scheme for the solar wind. J. Geophys. Res. Space Phys. **120**, 70–100 (2015)
18. Zhao, L., et al.: On the relation between the in-situ properties and the coronal sources of the solar wind. Astrophys. J. **846**(2), 135 (2017)
19. Zhao, L., Zurbuchen, T. H., Fisk, L.A.: Global distribution of the solar wind during solar cycle 23: ACE observations. GeoRL, **36** (2009)

An Improved Capsule Network for Speech Emotion Recognition

Huiyun Zhang[1,2] and Heming Huang[1,2(✉)]

[1] School of Computer Science, Qinghai Normal University, Xining, China
huanghm@qhnu.edu.cn
[2] The State Key Laboratory of Tibetan Intelligent Information Processing and Application,
Xining, China

Abstract. Speech Emotion Recognition (SER) is an important research content in the field of speech recognition, and it can make Human-Computer Interaction (HCI) more natural and harmonious. At present, the research of SER faces such problems as data scarcity and overfitting. To tackle these problems, an Improve Capsule Network (I-CapsNet) using Data Augmentation (DA) technique is proposed for efficient SER task. Firstly, the model I-CapsNet is used to reduce the number of parameters and retain more texture and background information. It can also retain the angle and discriminative information of the entity, thereby to relieve the overfiting. Secondly, noise is used to augment the baseline datasets (EMODB, CASIA, and SAVEE) under five Signal-to-Noise Ratios (SNR), to solve the problem of data scarcity. Finally, an evaluation index EAI is defined to comprehensively analyze the performance of the model under different data division namely EIRD, EDRD, EICV, and EDCV. Compared to the previous research results, the model I-CapsNet has competitive performance on the corresponding augmented datasets.

Keywords: Speech emotion recognition · Feature extraction · Capsule network · Data augmentation · Data scarcity

1 Introduction

As the fundamental research of affective computing, SER has become an active research area, and it provides users with a smoother interface and gives them reasonable feedback [1]. Dataset construction, feature extraction, and acoustic models are important research contents of SER. Therefore, building high-quality datasets, extracting the most representative features, and constructing robust and generalized models are the key to the success of SER [2].

The traditional models are Hidden Markov Models (HMM) [3], Gaussian Mixture Models (GMM) [4], Support Vector Machine (SVM) [5], and so on. HMM is a statistical probability model that can process time series. It is used to represent an observation sequence [6]. GMM can be regarded as a multi-dimensional Gaussian probability density function. It has been used in machine learning and other fields [7]. SVM first uses kernel to map feature vectors from input space to high-dimensional Hilbert space, and finds the optimal hyperplane to classify samples. However, it cannot solve the problem that the kernel matrix is too large due to large-scale training samples [8].

With the rise of deep learning, various Artificial Neural Networks (ANN) [9] have been introduced into acoustic modeling. Compared with traditional methods, these neural networks have better learning ability when dealing with large-scale data. However, different models have their own advantages and disadvantages. For example, Recurrent Neural Network (RNN) is good at processing time series information [10], and Convolutional Neural Network (CNN) is good at capturing spatial information [11].

Among the most advanced deep learning models, CNN-based models show strong performance. The structure of CNN is usually composed of the convolutional layer and the pooling layer. In which, Max-pooling layer, drops out the not 'max-information' inevitably. It also considers the information in its receptive field and extracts features in this local region. However, it ignores spatial relationships and orientation information in global region [11, 12]. To overcome these disadvantages, Sabour et al. [13] propose the Capsule Network (CapsNet), and it is introduced to different fields [14–16]. However, the complexity of CapsNet is high. Therefore, I-CapsNet, an Improved CapsNet (I-CapsNet) is proposed for the SER.

The model I-CapsNet needs a large amount of training data when carrying out emotion recognition tasks. However, most of the existing speech emotion datasets, such as CASIA [17], EMODB [18], and SAVEE [19], are relatively small. Therefore, these SER systems face the problem of data scarcity. The data scarcity problem may cause the deep learning models prone to overfitting, i.e., the model may only perform well on a few specific datasets, and it has poor generalization capability on other unseen data. Therefore, a Data Augmentation (DA) technique is used to augment datasets.

Specifically, the major contributions of this paper are summarized as: (1) The model I-CapsNet is proposed to detect emotions and decrease the overfitting risk. (2) The DA method is proposed to augment baseline datasets (EMODB, CASIA, and SAVEE) and to further decrease the overfitting.

2   The Proposed Methods

The scalars are transferred from lower-level neurons to upper-level neurons in CNN. Scalar has no direction and cannot represent the pose relationship between high-level features and low-level features. CNN has great limitations in identifying spatial relationship [11, 20, 21]. Therefore, Hinton et al. proposed the concept of CapsNet [13].

Compared with CNNs, the CapsNet uses a group of neurons instead of a unique neuron whose activity vector represents the instantiation parameters of a specific type of entity such as an object or object part. The CapsNet outputs a vector instead of a single scalar, which makes it can model the global spatial information and replace the pooling layers with dynamic routing algorithm to avoid losing the valuable information. The dynamic routing is an information selection mechanism to ensure that the outputs of child capsules are sent to proper parent capsules [22, 23].

To further improve the performance of the CapsNet, this study constructs an Improve Capsule Network, abbreviated as I-CapsNet, as shown in Fig. 1. The proposed model I-CapsNet consists of six modules. The first module is an input layer, the second module and the third module are composed of convolutional layer and maxpooling layer. The fourth module is composed of convolutional layer, the fifth module is composed of primary capsule layer, and the last module is digital capsule layer.

In the I-CapsNet model, 128 filters are used in the convolution layers, the convolution kernel size is set to 3*3, the pooling window size is set to 2*2, the primary capsule module is composed of 128 capsules, and each of which is a 8D vector. The number of digital capsule layer denotes the number of categories of emotion. The length of each digital capsule is 16 dimensions.

Each module in I-CapsNet has their own unique functionalities. The convolutional layer is adopted to ensure the sparsity of the network. The max-pooling layer is adopted to reduce the deviation of the estimated mean caused by the parameter error of the convolution layer and retain more texture information. The primary capsule layer is adopted to transform the feature maps extracted from the convolutional layer into vector capsules, and the digital capsule layer is used to classify emotions. There is a dynamic routing algorithm between the primary capsule layer and the digital capsule layer is used to carving out the most important part of the model. This makes the model more robust to the changes of target position and angle.

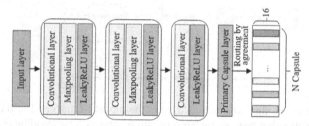

Fig. 1. The structure of the proposed model I-CapsNet

Figure 2 shows the dynamic routing procedure. In the previous section, the prediction vector u is computed through a weight matrix w. The dynamic routing determines the relationship between each parent capsule and its corresponding prediction vector. In this study, there are two iterative routing processes. The c_{ij} are the coupling coefficients, which is determined by a *Softmax*($\cdot$) function as shown in Eq. (1).

$$c_{ij} = \frac{\exp(b_{ij})}{\sum_{j} \exp(b_{ij})}$$

(1)

where the b_{ij} is the log prior probability that capsule i should be coupled to capsule j

$$\hat{u}_{j|i} = W_{ij} u_i \tag{2}$$

$$s_j = \sum c_{ij} \hat{u}_{j|i} \tag{3}$$

In Eq. (2) and Eq. (3), the prediction vectors, $\hat{u}_{j|i}$ is produced by multiplying a weight matrix W_{ij} by u_i which is the i-th output of a capsule in the $l-1$ layer. The total input to a capsule s_j is a weighted sum over all $\hat{u}_{j|i}$. In Eq. (4), the v_j is the vector output of capsule j in layer l, and s_j is its total input.

$$v_j = \frac{\|s_j\|}{1 + \|s_j\|} \frac{s_j}{\|s_j\|} \tag{4}$$

In the I-CapsNet, c_{ij}, s_j, and v_j are updated according to Eq. (1)–(4). To make b_{ij} more accurate, it is updated in multiple iterations according to Eq. (5).

$$b_{ij} = b_{ij} + \hat{u}_{j|i} \cdot v_j \tag{5}$$

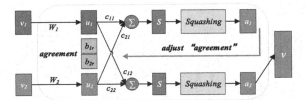

Fig. 2. The dynamic routing process employed in the model I-CapsNet. The routing algorithm determines the relationship between each parent capsule and its corresponding prediction vector.

3 Datasets and Feature Extraction

This section mainly describes the baseline datasets (EMODB, CASIA, and SAVEE), augmented datasets (single SNR noisy datasets (SSNDs), multiple SNRs noisy datasets (MSNDs)), and feature extraction procedure.

3.1 Baseline Datasets

To validate the effectiveness of the proposed model in SER, it has been tested on the three classic speech emotion datasets CASIA [17], EMODB [18], and SAVEE [19].

EMODB is a German dataset including 10 speakers and 7 emotions, i.e., boredom (B), anger (A), fear (F), sadness (Sa), disgust (D), happiness (H), and neutral (N), and it contains 535 emotional sentences in total.

CASIA is a Chinese dataset. The publicly CASIA speech emotion dataset contains 1200 utterances. There are 4 speakers and each speaker records 300 sentences in the same text. There are 6 emotions, i.e., anger (A), fear (F), happiness (H), neutral (N), sadness (Sa), surprise (Su).

SAVEE is an English dataset containing 4 speakers and 7 emotions, i.e., anger (A), disgust (D), fear (F), happiness (H), sadness (Sa), surprise (Su), and neutral (N). The sample number of neutral is 120 while that of each remainder class is 60. Totally, there are 480 utterances.

3.2 Augmented Datasets

To fully train the model I-CapsNet, the above baseline datasets are augmented by selecting 15 types of noises from Noisy dataset Noisex-92 [24] under 5 SNRs (-10, -5, 0, 5, and 10). The 15 kinds of noise are babble, buccaneer1, buccaneer2, volvo, f16, ml09, destroyerengine, destroyerops, factory1, factory2, hfchannel, leopard, pink, machinegun, and white. Therefore, to each baseline dataset, six augmented datasets are constructed, namely, five single SNR noisy datasets (SSND) and a multiple SNRs noisy datasets (MSND), as shown in Table 1. Where the MSND is constructed by merging five SSND. Thus, the number of samples in MSND is five times of that in SSND under the same baseline dataset.

Taking dataset CASIA as an example, 15 different types of noise are added under the five SNRs to obtain five augmented single-SNR noisy datasets, CASIA-10, CASIA-5, CASIA0, CASIA5, and CASIA10. Each single SNR dataset contains 18000 emotion samples. They are further merged to obtain a new multiple SNRs dataset CASIAM with 90000 samples. The other two baseline datasets, namely EMODB and SAVEE, are augmented and merged exactly in the same way.

3.3 Feature Extraction

The input of the proposed model I-CapsNet consists of 21D LLD (Low-level Descriptor) features [25]. Each speech is segmented into frames with a 25 ms window and 10 ms shifting step. To each frame, LLD features, including 1D ZCR and 20D MFCC. Some features of the samples are quite different. When the samples are directly input into the training, the contour of the loss function will be flat and long. Before finding the optimal solution, the gradient descent process is not only tortuous but very time-consuming. Therefore, it is necessary to quantify each feature into a unified interval.

After data standardization, all indicators are in the same order of magnitude, and they are suitable for comprehensive evaluation. After feature normalization, the contour of the loss function will be partial circle, the gradient descent process will be flatter [26]. Furthermore, the convergence will be faster, and therefore, the performance will be improved. We normalized the dataset so that the new dataset has a zero mean and unit variance [27]. It should be noted that the normalization procedure should be applied to each feature rather than each sample.

Table 1. The specific information about SSND and MSND constructed according to the baseline datasets (EMODB, CASIA, and SAVEE)

Datasets	Noise	SSND	Emotions/samples	Total samples	MSND	Emotions/samples	Total samples
CASIA	babble, white, buccaneer1, buccaneer2, destroyerengine, destroyerops, f16, factory1, factory2, volvo, hfchannel, pink, leopard, ml09, Machinegun	CASIA-10	A/3000, F/3000, H/3000, Sa/3000, Su/3000, N/3000	18000	CASIAM	A/15000, F/15000, H/15000, Sa/15000, Su/15000, N/15000	90000
		CASIA-5		18000			
		CASIA0		18000			
		CASIA5		18000			
		CASIA10		18000			
EMODB		EMODB-10	A/1905, B/1215, D/690, H/1065, F/1035, Sa/930, N/1185	8025	EMODBM	A/9525, B/6075, D/3450, H/5325, F/5175, Sa/4650, N/5925	40125
		EMODB-5		8025			
		EMODB0		8025			
		EMODB5		8025			
		EMODB10		8025			
SAVEE		SAVEE-10	A/900, D/900, F/900, H/900, Sa/900, Su/900, N/1800	7200	SAVEEM	A/4500, D/4500, F/4500, H/4500, Sa/4500, Su/4500, N/9000	36000
		SAVEE-5		7200			
		SAVEE0		7200			
		SAVEE5		7200			
		SAVEE10		7200			

4   Experiments and Analysis

This section mainly includes the experimental setup, model parameter configuration, data division methods, and results analysis.

4.1   Experimental Setup

All experiments are performed on a powerful PC with 64G RAM running under Windows 10. CPU speed is 2.10 GHz, core is 40, and logic processor is 80. 2 RTX 2080 Ti GPUs are used to accelerate the speed of computation. All models are implemented with TensorFlow toolkit [28].

4.2   Model Parameter Configuration

The parameter configurations of the model I-CapsNet are as follows: 128 filters are used in convolution layers, the convolution kernel size is set to 3*3, and the step size is 1, the pooling window size is set to 2*2, the activation function is LeakyReLU [29], and the optimizer is Adam [30]. When the emotion category is 7, the number of parameters of the model is 310784.

4.3   Data Division Methods

Data division methods, namely Emotion-Independent Random-Division (EIRD), Emotion-Dependent Random-Division (EDRD), Emotion-Independent Cross-Validation (EICV), and Emotion-Dependent Cross-Validation (EDCV), are adopted to evaluate the performance of the model I-CapsNet.

In the EIRD method, all samples are randomly divided into five equal parts, and four parts are selected as the training data while the remaining one part as the test data. In the EDRD method, all samples of each emotion category are divided into five equal parts. To each part, each part is randomly selected as the test data, and the remaining samples are used as training data. In the EICV method, all samples are randomly divided into five parts, and each part is used as the test data in turn while the remaining four parts are used as the training data. In the EDCV method, samples of each emotion category are randomly divided into five parts. Each part is used as the test data in turn while the remaining four parts are used as the training data. The EICV and EDCV can ensure that all samples are used in both the training and test.

To evaluate the robustness and generalization of the model more comprehensively, a new evaluation index Expected Accuracy Index (EAI) is introduced, and it is defined as the expected accuracy from different training and test data partitions.

$$\text{EAI} = \frac{1}{N}(acc_{EIRD} + acc_{EDRD} + acc_{EICV5} + acc_{EDCV5})$$

where acc_{EIRD}, acc_{EDRD}, acc_{EDRD}, and acc_{EDRD} denote the accuracy of the model I-CapsNet under the data division methods EIRD, EDRD, EICV, and EDCV, respectively.

4.4 Experimental Results and Analysis

We use four data division methods to verify the performance of the model on the baseline datasets, SSND, and MSND, respectively.

4.4.1 Results on the Baseline Datasets

The robustness and generalization of the model I-CapsNet are verified on the baseline datasets, as shown in Table 2. It can be seen from Table 2:

Table 2. The results of the model I-CapsNet on the baseline datasets under four data division methods, namely, EDCV, EICV, EDRD, and EIRD

Data division	Folds	CASIA					
		1	2	3	4	5	Avg ± Var
EDCV	K_folds = 5	76.67	72.50	73.33	72.92	75.00	74.08 ± 2.39
EICV	K_folds = 5	76.25	75.83	77.50	71.25	77.08	75.58 ± 5.04
EDRD	5 times	74.58	79.17	77.50	72.50	75.83	75.92 ± 5.31
EIRD	5 times	75.83	72.50	73.33	72.08	74.17	73.58 ± 1.78
Data division	Folds	EMODB					
		1	2	3	4	5	Avg ± Var
EDCV	K_folds = 5	78.50	80.37	85.98	77.57	77.57	80.00 ± 9.99
EICV	K_folds = 5	79.44	81.31	83.18	71.96	77.57	78.69 ± 14.83
EDRD	5 times	80.37	73.83	77.57	79.44	77.57	77.76 ± 5.03
EIRD	5 times	73.83	72.90	73.83	75.70	71.96	73.64 ± 1.54
Data division	Folds	SAVEE					
		1	2	3	4	5	Avg ± Var
EDCV	K_folds = 5	65.62	62.50	62.50	55.21	55.21	60.21 ± 17.95
EICV	K_folds = 5	63.54	66.67	58.33	59.38	56.25	60.83 ± 14.16
EDRD	5 times	55.21	60.42	58.33	58.33	58.33	58.12 ± 2.78
EIRD	5 times	54.17	55.21	53.12	59.38	57.29	55.83 ± 5.04

(1) The performance of the model is easily affected by the data division methods, which is mainly reflected in two aspects. First, the performance of the model is sensitive to the data division. Second, the results of each experiment are quite different under the same data division.

(2) There is no absolute difference between the four data division methods. For example, the performance of the model reaches the highest (75.92%) on the CASIA dataset with EDRD. However, the performance of the model reaches the highest on the EMODB dataset with EDCV and on the SAVEE dataset with EICV.

To more intuitively show the overall performance of the model under the data division methods EDCV, EICV, EDRD, and EIRD. Figure 3 illustrates the violin diagrams to represent the results of the model. In the violin diagram, the blue five-pointed star represents the median. The following conclusions can be drawn:

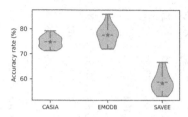

Fig. 3. The performance of model I-CapsNet on the datasets CASIA, EMODB, and SAVEE

It can be seen that the results of the model I-CapsNet on dataset EMODB is the best, and that on dataset SAVEE is the worst. The main reason is that the SAVEE contains seven types of emotions and fewer samples. The model parameters better fit the data distribution on the dataset EMODB. In addition, the robustness of the model on the dataset CASIA is appreciable, mainly because the number of samples on the dataset CASIA is larger. Comparatively speaking, and the model can be fully trained and captured emotion related features.

4.4.2 Results on the SSND and MSND

Due to the small number of samples on the baseline datasets, the model cannot be fully trained. For this reason, we have verified the effectiveness of the I-CapsNet on the augmented datasets. Tables 3, 4 and 5 show the results of the model on the SSNDs and MSNDs. The following conclusions can be drawn:

(1) The results of the I-CapsNet model on the SSND and MSND are affected by the different data division. It shows the same change trend as that on the corresponding datasets. The stability on the SSND and MSND is higher than that on the baseline datasets under the different data division methods. However, no data division method has the best performance on all datasets, as shown in Fig. 4(a). The model achieves high performance on the CASIA and EMODB datasets with EDCV method under the SNR $= 0$. The model achieves high performance with EICV on the SAVEE dataset.

(2) The performance of the model is positively correlated with SNRs, namely, the performance of the model will improve linearly as the SNR increases from -10 to 10. For example, when the EDCV method is used, the accuracy of the model is 56.89% with SNR $= -10$, and 99.26% with SNR $= 10$ on the EMODB dataset.

(3) The results on the MSNDs (EMODBM, CASIAM, and SAVEEM) can more comprehensively and objectively represent the impact of SNR on the model performance, as shown in Fig. 4(b). First, the accuracy of the model are greater than 80.44% (SAVEEM with EDRD) on all MSNDs, which is greatly improved from

Table 3. The performance of the I-CapsNet model on the SSNDs (EMODB-10, EMODB-5, EMODB0, EMODB5, and EMODB10) and MSND (EMODBM) datasets

Data sivision	Folds	SSND(SNR = −10)/EMODB-10						SSND(SNR = −5)/EMODB-5					
		1	2	3	4	5	Avg ± Var	1	2	3	4	5	Avg ± Var
EDCV	5 times	56.76	56.95	56.95	56.64	57.13	56.89 ± 0.03	74.14	73.02	67.66	71.84	71.84	71.70 ± 4.81
EICV	5 times	57.26	58.69	58.13	56.88	56.01	57.39 ± 0.88	73.33	70.84	70.59	71.59	69.35	71.14 ± 1.72
EDRD	5 times	56.95	56.64	57.01	58.32	58.50	57.48 ± 0.59	73.08	73.96	71.78	73.15	73.27	73.05 ± 0.50
EIRD	5 times	58.07	57.82	55.58	59.25	57.69	57.68 ± 1.41	72.15	72.83	72.40	73.64	72.06	72.62 ± 0.33

Data division	Folds	SSND(SNR = 0)/EMODB0						SSND(SNR = 5)/EMODB5					
		1	2	3	4	5	Avg ± Var	1	2	3	4	5	Avg ± Var
EDCV	5 times	86.54	89.72	88.22	89.60	87.35	88.29 ± 1.54	96.07	96.76	96.32	96.51	96.20	96.37 ± 0.06
EICV	5 times	87.60	87.73	87.29	90.03	85.86	87.70 ± 1.80	97.32	97.01	97.26	95.83	97.57	97.00 ± 0.37
EDRD	5 times	86.79	86.92	88.29	87.23	86.04	87.05 ± 0.53	96.14	96.07	94.83	95.89	95.08	95.60 ± 0.29
EIRD	5 times	87.91	87.98	88.29	88.10	88.10	88.08 ± 0.02	96.14	95.76	96.88	96.45	95.95	96.24 ± 0.16

Data division	Folds	SSND(SNR = 10)/EMODB10						MSND(Multi-SNR5)/EMODBM					
		1	2	3	4	5	Avg ± Var	1	2	3	4	5	Avg ± Var
EDCV	5 times	99.44	98.88	99.07	99.31	99.44	99.23 ± 0.05	89.32	89.89	88.90	88.87	88.05	89.01 ± 0.36
EICV	5 times	98.69	99.38	99.31	98.82	99.50	99.14 ± 0.10	89.37	89.46	89.26	89.50	88.91	89.30 ± 0.04
EDRD	5 times	98.94	99.56	99.50	99.25	99.56	99.36 ± 0.06	90.12	89.43	89.63	89.05	89.01	89.45 ± 0.17
EIRD	5 times	98.75	99.38	99.44	99.31	99.44	99.26 ± 0.07	89.05	90.36	89.46	89.53	88.85	89.45 ± 0.28

Table 4. The performance of the proposed I-CapsNet model on the SSNDs (CASIA-10, CASIA-5, CASIA0, CASIA5, and CASIA10) and MSND (CASIAM) datasets

Data division	Folds	SSND(SNR = −10)/CASIA-10						SSND(SNR = −5)/CASIA-5					
		1	2	3	4	5	Avg ± Var	1	2	3	4	5	Avg ± Var
EDCV	5 times	56.47	57.33	59.86	55.89	57.22	57.35 + 1.84	66.86	67.22	66.56	64.97	67.36	66.59 + 0.74
EICV	5 times	58.00	58.11	57.06	55.89	57.75	57.36 + 0.68	65.94	69.17	68.08	69.50	68.64	68.27 + 1.58
EDRD	5 times	58.68	57.97	57.83	57.69	56.86	57.81 + 0.34	65.08	68.25	68.25	68.08	68.06	67.54 + 1.52
EIRD	5 times	57.64	56.25	56.89	56.89	57.89	57.11 + 0.35	68.69	69.11	69.31	69.06	69.08	69.05 + 0.04

Data division	Folds	SSND(SNR = 0)/CASIA0						SSND(SNR = 5)/CASIA5					
		1	2	3	4	5	Avg ± Var	1	2	3	4	5	Avg ± Var
EDCV	5 times	82.00	80.97	82.14	83.00	80.58	81.74 + 0.75	92.19	92.69	90.89	91.00	90.25	91.40 + 0.81
EICV	5 times	80.67	81.42	80.56	82.50	80.67	81.16 + 0.54	91.42	92.25	93.25	91.06	90.83	91.76 + 0.79
EDRD	5 times	81.50	79.56	81.83	81.19	82.36	81.29 + 0.90	91.39	93.56	91.97	92.08	92.25	92.25 + 0.51
EIRD	5 times	81.97	81.75	79.86	81.42	81.42	81.28 + 0.55	92.42	91.92	93.19	92.36	90.56	92.09 + 0.75

Data division	Folds	SSND(SNR = 10)/CASIA10						MSND(Multi-SNR5)/CASIAM					
		1	2	3	4	5	Avg ± Var	1	2	3	4	5	Avg ± Var
EDCV	5 times	96.89	96.81	97.50	96.19	97.31	96.94 + 0.21	87.79	87.41	86.60	86.89	87.09	87.16 + 0.17
EICV	5 times	96.72	96.86	95.78	96.81	96.67	96.57 + 0.16	87.24	86.86	87.85	85.91	86.27	86.83 + 0.47
EDRD	5 times	97.72	96.58	96.75	97.03	96.75	96.97 + 0.16	87.13	86.13	87.20	87.34	86.32	86.82 + 0.25
EIRD	5 times	97.28	97.03	96.53	97.03	97.58	97.09 + 0.12	87.02	87.57	86.59	86.71	86.77	86.93 + 0.12

Table 5. The performance of the proposed I-CapsNet model on the SSNDs (SAVEE-10, SAVEE-5, SAVEE0, SAVEE5, and SAVEE10) and MSND (SAVEEM) datasets

Data division	Folds	SSND(SNR = -10)/SAVEE-10						SSND(SNR = -5)/SAVEE-5					
		1	2	3	4	5	Avg ± Var	1	2	3	4	5	Avg ± Var
EDCV	5 times	45.21	47.57	46.46	48.40	46.81	46.89 ± 1.15	57.43	60.28	57.01	56.11	57.36	57.64 ± 1.97
EICV	5 times	48.68	46.94	46.81	46.32	46.60	47.07 ± 0.69	58.40	56.60	56.88	56.39	59.72	57.60 ± 1.62
EDRD	5 times	47.01	48.75	47.29	47.43	46.81	47.46 ± 0.46	57.78	57.85	57.50	57.22	59.24	57.92 ± 0.49
EIRD	5 times	46.18	48.68	48.68	50.14	49.24	48.58 ± 1.73	59.03	56.53	58.47	58.26	58.47	58.15 ± 0.72

Data division	Folds	SSND(SNR = 0)/SAVEE0						SSND(SNR = 5)/SAVEE5					
		1	2	3	4	5	Avg ± Var	1	2	3	4	5	Avg ± Var
EDCV	5 times	74.37	74.31	71.74	72.64	72.43	73.10 ± 1.12	89.44	88.68	85.69	87.43	86.39	87.53 ± 1.93
EICV	5 times	75.90	73.40	74.51	73.26	74.17	74.25 ± 0.90	88.68	88.33	85.63	89.17	87.92	87.95 ± 1.51
EDRD	5 times	72.71	72.92	71.32	73.06	73.47	72.70 ± 0.54	87.36	86.88	89.44	86.04	88.75	87.69 ± 1.53
EIRD	5 times	73.06	74.93	73.40	74.93	74.31	74.13 ± 0.60	87.57	88.40	87.92	88.47	88.89	88.25 ± 0.21

Data division	Folds	SSND(SNR = 10)/SAVEE10						MSND(Multi-SNR5)/SAVEEM					
		1	2	3	4	5	Avg ± Var	1	2	3	4	5	Avg ± Var
EDCV	5 times	95.76	96.32	95.56	96.04	95.90	95.92 ± 0.07	80.51	81.28	81.61	81.19	81.24	81.17 ± 0.13
EICV	5 times	95.83	96.18	95.97	96.04	95.35	95.87 ± 0.08	80.71	80.93	80.67	80.10	81.44	80.77 ± 0.19
EDRD	5 times	95.69	94.44	95.76	96.67	96.04	95.72 ± 0.53	80.26	81.07	80.54	79.81	80.53	80.44 ± 0.17
EIRD	5 times	96.18	95.97	95.97	96.04	96.88	96.21 ± 0.12	81.57	81.22	81.46	81.03	80.81	81.22 ± 0.08

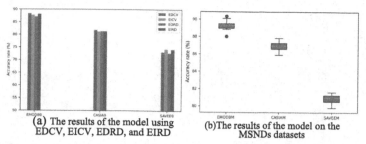

(a) The results of the model using EDCV, EICV, EDRD, and EIRD

(b)The results of the model on the MSNDs datasets

Fig. 4. The results of the model I-CapsNet. (a) The performance of the model using EDCV, EICV, EDRD, and EIRD methods under the SNR = 0. (b) The performance of the model on the EMODBM, CASIAM, and SAVEEM

that on the baseline datasets. It indicates that the MSNDs are very profound significance. Second, the result on the dataset EMODBM is the best, mainly because it is trained fully on the dataset EMODBM, and the parameters fit better the data distribution on the dataset EMODBM.

(4) The I-CapsNet model has a good accuracy of all emotions on the dataset EMODB10, CASIA10, and SAVEE10, but there are still some differences in the recall of various emotions, as shown in Fig. 5. Specifically, the recall of the model on datasets EMODB10, CASIA10, and SAVEE10 are 99.29%, 96.14%, and 94.84%, respectively. Among them, on the EMODB10 dataset, the recall of 'boredom' emotion is relatively low, because there are relatively few samples. On the CASIA10 and SAVEE10 dataset, the recall of 'anger' emotion is the highest. This is mainly because its samples have higher energy in arousal, valence, and dominance dimension, and therefore they are easier to distinguish.

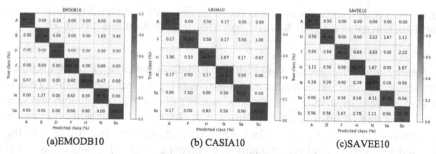

(a)EMODB10   (b) CASIA10   (c)SAVEE10

Fig. 5. Confusion matrices of the proposed model I-CapsNet on the EMODB10, CASIA10, and SAVEE10 datasets

(5) In t-SNE visualizations, the distinctiveness between the same emotions is small, and the distinctiveness between different emotions is large, as shown in Fig. 6.

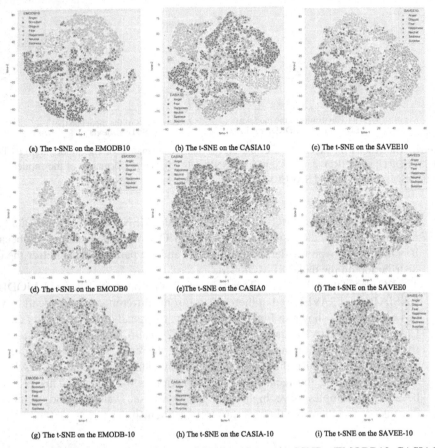

(a) The t-SNE on the EMODB10   (b) The t-SNE on the CASIA10   (c) The t-SNE on the SAVEE10

(d) The t-SNE on the EMODB0   (e)The t-SNE on the CASIA0   (f) The t-SNE on the SAVEE0

(g) The t-SNE on the EMODB-10   (h) The t-SNE on the CASIA-10   (i) The t-SNE on the SAVEE-10

Fig. 6. The t-SNE visualizations of the model I-CapsNet on the SSNDs (EMODB10, CASIA10, SAVEE10, EMODB0, CASIA0, SAVEE0, EMODB-10, CASIA-10, and SAVEE-10)

Among them, the separability of EMODB10 dataset is the best. Except that boredom and sadness are confused with other emotions to varying degrees, the remaining emotions are well distinguished, mainly because the number of samples of boredom and sadness is less than that of other emotions. On the CASIA dataset, fear and other emotions are confused to varying degrees. In CASIA dataset, sadness and disgust are highly confused with other emotions. In addition, comparing (a)(d)(g), (b)(e)(h), and (c)(f)(i) respectively, when the SNR is 10, the separability of the model is very good, and the all kinds of emotions can be well distinguished; When the SNR is 0, although there is a certain degree of confusion among the different emotions, they are still separable. When the SNR = −10, almost all emotion categories are confused, and the separability is poor. It indicates that the positive SNR can promote emotion classification, while the negative SNR will inhibit emotion classification.

4.4.3   Compared with the Previous Results

Table 6 shows the performance of the proposed model on the baseline datasets (EMODB, CASIA, and SAVEE) and its corresponding MSNDs, namely CASIAM, EMODBM, and SAVEEM, under four data division methods EIRD, EDRD, EICV, and EDCV, as can be seen from Table 6:

Table 6. Performance of the model I-CapsNet on baseline datasets and MSND. The EAI is used to comprehensively measure the performance of the model

Datasets		EIRD	EDRD	EICV	EDCV	EAI + Std
Baseline datasets	CASIA	73.58	75.92	75.58	74.08	74.79 + 0.97
	EMODB	73.64	77.76	78.69	80.00	77.52 + 5.66
	SAVEE	55.83	58.12	60.83	60.21	58.75 + 3.85
MSND	CASIAM	86.93	86.82	86.83	87.16	86.94 + 0.02
	EMODBM	89.45	89.45	89.30	89.01	89.30 + 0.03
	SAVEEM	81.22	80.44	80.77	81.17	80.90 + 0.10

Compared with the baseline datasets, no matter which data division is used, the I-CapsNet has achieved considerable recognition results on the corresponding MSNDs, and it shows that it is feasible to carry out SER researches by augmenting datasets.

Among three MSNDs, the model I-CapsNet has the best performance on the dataset EMODBM, with the EAI reached 89.30%. The performance on the dataset SAVEEM is the lowest, with the EAI is 80.90%. The main reasons are the small number of samples contained on the dataset SAVEEM and the large number of emotion categories.

The stability of the model on the MSNDs are much higher than that on the corresponding baseline datasets. The main reason is that the MSNDs have more samples, therefore, the model is more fully trained and the robustness is also improved.

Finally, we validate the effect of different SNRs on the model performance. Figure 7 shows the performance of the model I-CapsNet under different SNRs on the datasets CASIA, EMODB, and SAVEE. Among them, baseline refers to the results of the model without DA. Multi-SNR-5 refers to the performance of the model on the MSNDs.

As can be seen from Fig. 7 no matter on which dataset, the performance of the model I-CapsNet is the worst when the SNR is -10. As the SNR gradually changes from −10 to 10, the performance of the model increases linearly, that is, when the SNR is 10, the performance of the model reaches the maximum. In addition, the augmentation datasets with positive SNRs (SNR > = 0) can promote the model performance, while the augmentation datasets with negative SNR data will inhibit the model performance. When the baseline datasets are augmented with five SNRs, more objective performance of the model I-CapsNet is achieved.

To further highlight the advantages of the proposed model. Table 7 summarizes the performance improvement of the proposed model I-CapsNet to the related peer methods on baseline datasets CASIA, EMODB, and SAVEE.

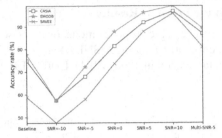

Fig. 7. The performance comparison of the model I-CapsNet under the different SNRs

Table 7. The performance of the model I-CapsNet to those of the peers in literature on baseline datasets CASIA, EMODB, and SAVEE

Datasets	Model	UAR	Datasets	Model	UAR	Datasets	Model	UAR
CASIA	CapCNN [31]	81.90	EMODB	ACRNN [34]	82.82	SAVEE	FFBPANN [36]	71.70
	FDNNSA [32]	82.08		AFSS [18]	83.00		RBFNN [37]	72.96
	LSTM-TF-at [33]	92.80		WADAN [35]	83.31		SVM [36]	77.32
	I-CapsNet (Ours)	86.94		I-CapsNet (Ours)	89.30		I-CapsNet (Ours)	80.90

For dataset CASIA, Wen et al. proposed a CapCNN model that combines CNN and CapsNet, which achieves an accuracy of 81.90% [31]. Based on human emotions and identification information, Chen et al. proposed a FDNNSA (Fuzzy Deep Neural Network with Sparse Autoencoder) for intention understanding and it achieves an accuracy of 82.08% [32]. Xie et al. proposed a LSTM-TF-Att (Attention-based LSTM model on Time and Frequency dimensional), and it achieves an accuracy of 92.80% [33].

For dataset EMODB, Chen et al. proposed a 3D ACRNN (Attention-based Convolution Recurrent Neural Networks) to learn discriminative features for SER. It achieves an accuracy of 82.82% [34]. Cirakman et al. introduced an AFSS (Active Field State Space) method which incorporates particle filter tracking for switching observation models with emotion classification, and it achieves an accuracy of 83.00% [18]. Yi et al. proposed a new WADAN (Adversarial Data Augmentation Network with Wasserstein Divergence), which consists of a GAN, an autoencoder, and an auxiliary classifier. The proposed model achieves an accuracy of 83.31%, and it is competitive with state-of-the-art SER systems [35].

For dataset SAVEE, Sugan et al. utilized the FF-BP-ANN (Feedforward Backpropagation Artificial Neural Network) and SVM to recognize emotions and their accuracies are 71.70% and 77.32%, respectively [36]. Panigrahi et al. proposed a RBFNN (Radial Basis Function Neural Network), and it achieves an accuracy of 72.96% [37].

It can be seen from Table 7 that, except on dataset CASIA, the performances of the I-CapsNet model on EMODB and SAVEE are better than the previous results, therefore, it is safe to say that the model I-CapsNet is more effective. In addition, the feature dimension is low, the model parameters are less, and the model training speed is fast. Therefore, it meets the requirements of real-time system.

5 Conclusion

In this work, I-CapsNet with DA technique, is proposed to tackle the data scarcity and decrease the overfitting problems. The I-CapsNet model consists of input layer, convolutional layer, maxpooling layer, primary capsule layer, and digital capsule layer. The convolutional layer is used to extract different input features. The lower convolutional layer can only extract some low-level features, such as edges, lines, and angles. The higher convolutional layer can iteratively extract more complex features from the low-level features. The max-pooling layer is used to reduce the number of parameters and retain more texture and background information. The primary capsule layer is adopted to transform the feature maps extracted from the convolution layer into vector capsules, and the digital capsule layer is used to classify emotions. The dynamic routing algorithm is used to carve out the most important information. Besides, we used 15 types of noise to augment the baseline datasets. DA technique is developed to augment the number of samples and make the model more fully trained. Compared to the existing approaches, the I-CapsNet achieves more competitive performance.

The directions that can be further explored in the future are as follows: (1) Try to apply the I-CapsNet to conduct cross-domain SER research. (2) Try to extract the different speech emotion features and seek the optimal feature.

Acknowledgements. The authors acknowledge the National Natural Science Foundation of China (Grant: 62066039) and Natural Science Foundation of Qinghai Province (Grant: 2022-ZJ-925).

References

1. Ramakrishnan, S., Emary, I.: Speech emotion recognition approaches in human computer interaction. Telecommun. Syst. **52**, 1467–1478 (2013)
2. John, K., Saurous, R.: Emotion recognition from human speech using temporal information and deep learning. In: Proceedings of Interspeech, Hyderabad, India, pp. 937–940 (2018)
3. Mao, S., Tao, D., Zhang, G., Ching, P.C., Lee, T.: Revisiting hidden Markov models for speech emotion recognition. In: Proceedings of IEEE International Conference on Acoustics, Speech and Signal Processing (ICASSP), Brighton, United Kingdom, pp. 6715–6719 (2019)
4. Shahin, I., Nassif, A.B., Hamsa, S.: Emotion recognition using hybrid Gaussian mixture model and deep neural network. IEEE Access **7**, 26777–26787 (2019)
5. Teng, Z., Ren, F., Kuroiwa, S.: Emotion recognition from text based on the rough set theory and the support vector machines. In: Proceedings of Natural Language Processing and Knowledge Engineering, Beijing, China, pp. 36–41 (2007)

6. Song, M., Chen, C., You, M.: Audio-visual based emotion recognition using tripled hidden Markov model. In: Proceedings of IEEE International Conference on Acoustics, Speech and Signal Processing (ICASSP), Montreal, Canada, pp. 877–880 (2004)
7. Vydana, H.K., Kumar, P.P., Krishna, K.S.R., Vuppala, A.K.: Improved emotion recognition using GMM-UBMs. In: Proceedings of IEEE International Conference on Signal Processing and Communication Engineering Systems, Guntur, India, pp. 53–57 (2015)
8. Hu, H., Xu, M.-X., Wu, W.: GMM supervector based SVM with spectral features for speech emotion recognition. In: Proceedings of IEEE International Conference on Acoustics, Speech and Signal Processing (ICASSP), Honolulu, USA, pp. 413–416 (2007)
9. Mao, X., Chen, L., Fu, L.: Multi-level speech emotion recognition based on HMM and ANN. In: Proceedings of WRI World Congress on Computer Science and Information Engineering, Los Angeles, USA, pp. 225–229 (2009)
10. Chen, X., Han, W., Ruan, H., et al.: Sequence-to-sequence modelling for categorical speech emotion recognition using recurrent neural network. In: Proceedings of First Asian Conference on Affective Computing and Intelligent Interaction, Beijing, China, pp. 1–4 (2018)
11. Bertero, D., Fung, P.: A first look into a convolutional neural network for speech emotion detection. In: Proceedings of IEEE International Conference on Acoustics, Speech and Signal Processing (ICASSP), New Orleans, USA, pp. 5115–5119 (2017)
12. Li, R., Wu, Z., Jia, J., Zhao, S., Meng, H.: Dilated residual network with multi-head self-attention for speech emotion recognition. In: Proceedings of IEEE International Conference on Acoustics, Speech and Signal Processing (ICASSP), Brighton, UK, 2019, pp. 6675–6679
13. Sabour, S., Frosst, N., Hinton, G.E.: Dynamic routing between capsules. In: Proceedings of Advances in Neural Information Processing Systems, 2017, pp. 3856–3866.
14. Duarte, K., Rawat, Y.S., Shah, M.: Videocapsulenet: a simplified network for action detection. In: Proceedings of Advances in Neural Information Processing Systems (NIPS), Long Beach, US, pp. 7610–7619 (2018)
15. Zhang, B.W., Xu, X.F., Yang, M., Chen, X.J., Ye, Y.M.: Cross-domain sentiment classification by capsule network with semantic rules. IEEE Access **6**, 1–1 (2018)
16. Min, Y., Zhao, M., Ye, J.B., Lei, Z., Zhang, S.: Investigating capsule networks with dynamic routing for text classification. In: Proceedings of the Conference on Empirical Methods in Natural Language Processing (EMNLP), Brussels, Belgium, pp. 3110–3119 (2018)
17. Wang, K., An, N., Li, B.N., Zhang, Y., Li, L.: Speech emotion recognition using Fourier parameters. IEEE Trans. Affect. Comput. **25**, 69–75 (2015)
18. Cirakman, O., Gunsel, B.: Online speaker emotion tracking with a dynamic state transition model. In: Proceedings of International Conference on Pattern Recognition (ICPR), Cancun, Mexico, pp. 307–312 (2016)
19. Kim, Y., Provost, E.: ISLA: Temporal segmentation and labeling for audio-visual emotion recognition. IEEE Trans. Affect. Comput. **10**(2), 196–208 (2019)
20. George, T., Fabien, R., Raymond, B.: Adieu features? end-to-end speech emotion recognition using a deep convolution recurrent network. In: Proceedings of IEEE International Conference on Acoustics, Speech and Signal Processing, Shanghai, China, pp. 5200–5204 (2016)
21. Wang, L., Dang, J., Zhang, L.: Speech emotion recognition by combining amplitude and phase information using convolutional neural network. In: Proceedings of Interspeech, Hyderabad, India, pp. 1611–1615 (2018)
22. Wu, X.X., Liu, S.X., Cao, Y.W., Li, X., Yu, J.W., Dai, D.Y.: Speech emotion recognition using capsule network. In: Proceedings of IEEE International Conference on Acoustics, Speech and Signal Processing (ICASSP), Brighton, UK, pp. 6695–6699 (2019)
23. Xiang, C.Q., Zhang, L., Tang, Y., Zou, W.B., Xu, C.: MS-CapsNet: a novel multi-scale capsule network. IEEE Signal Process. Lett. **25**(12), 1850–1854 (2018)
24. Janovi, P., Zou, X.: Speech enhancement based on Sparse Code Shrinkage employing multiple speech models. Speech Commun. **54**(1), 108–118 (2012)

25. Gideon, J., McInnis, M.G., Provost, E.M.: Improving cross-corpus speech emotion recognition with adversarial discriminative domain generalization (ADDoG). IEEE Trans. Affect. Comput. **12**(4), 1055–1068 (2021)
26. Pappagari, R., Villalba, J., Želasko, P.: CopyPaste: An augmentation method for speech emotion recognition. In: Proceedings of IEEE International Conference on Acoustics, Speech and Signal Processing (ICASSP), Toronto, Canada, pp. 6324–6328 (2021)
27. Raju, V.N.G., Lakshmi, K.P., Jain, V.M., Kalidindi, A., Padma, V.: Study the influence of normalization/transformation process on the accuracy of supervised classification. In: Proceedings of International Conference on Smart Systems and Inventive Technology (ICSSIT), Tirunelveli, India, pp. 729–735 (2020)
28. Ertam, F., Aydın, G.: Data classification with deep learning using Tensorflow. In: Proceedings of International Conference on Computer Science and Engineering (UBMK), Antalya, Turkey, pp. 755–758 (2017)
29. Jiang, T., Cheng, J.: Target recognition based on CNN with LeakyReLU and PReLU activation functions. In: Proceedings of International Conference on Sensing, Diagnostics, Prognostics, and Control (SDPC), Beijing, China, pp. 718–722 (2019)
30. Chen, K., Ding, H., Huo, Q.: Parallelizing Adam optimizer with blockwise model-update filtering. In: Proceedings of IEEE International Conference on Acoustics, Speech and Signal Processing (ICASSP), Barcelona, Spain, pp. 3027–3031 (2020)
31. Wen, X.C., Liu, K.H., Zhang, W.M., Jiang, K.: The application of capsule neural network-based CNN for speech emotion recognition. In: Proceedings of International Conference on Pattern Recognition (ICPR), Milan, Italy, pp. 9356–9362 (2021)
32. Chen, L., Su, W., Wu, M., Pedrycz, W., Hirota, K.: A fuzzy deep neural network with sparse autoencoder for emotional intention understanding in human-robot interaction. IEEE Trans. Fuzzy Syst. **28**(7), 1252–1264 (2020)
33. Xie, Y., Liang, R., Liang, Z., Huang, C., Zou, C., Schuller, B.: Speech emotion classification using attention-based LSTM. In: IEEE/ACM Transactions on Audio, Speech, and Language Processing, vol. 27, no. 11, pp. 1675–1685 (2019)
34. Chen, M., He, X., Yang, J., Zhang, H.: 3-D convolutional recurrent neural networks with attention model for speech emotion recognition. IEEE Signal Process. Lett. **25**(10), 1440–1444 (2018)
35. Yi, L., Mak, M.W.: Improving speech emotion recognition with adversarial data augmentation network. IEEE Trans. Neural Netw. Learn. Syst. **33**(1), 172–1844 (2022)
36. Sugan, N., Sai Srinivas, N.S., Kar, N., Kumar, L.S., Nath, M.K., Kanhe, A.: Performance comparison of different cepstral features for speech emotion recognition. In: Proceedings of International CET Conference on Control, Communication, and Computing (IC4), Thiruvananthapuram, India, pp. 266–271 (2018)
37. Panigrahi, S.N., Palo, H.K.: Emotional speech recognition using particle swarm optimization algorithm. In: Proceedings of International Conference in Advances in Power, Signal, and Information Technology (APSIT), Bhubaneswar, India, pp. 1–5 (2021)

Distributed Query Processing and Reasoning Over Linked Big Data

Hamza Haruna Mohammed[1,3] [ID], Erdogan Doğdu[2(✉)] [ID], Roya Choupani[2] [ID], and Tomiya S. A. Zarbega[4]

[1] Norwegian University of Science and Technology, Trondheim, Norway
hamza.mohammed@ntnu.no
[2] Angelo State University, San Angelo, TX, USA
{erdogan.dogdu, roya.choupani}@angelo.edu
[3] Çankaya University, Ankara, Turkey
[4] Kastamonu University, Kastamonu, Turkey

Abstract. The enormous amount of structured and unstructured data on the web and the need to extract and derive useful knowledge from this big data make Semantic Web and Big Data Technology explorations of paramount importance. Open semantic web data created using standard protocols (RDF, RDFS, OWL) consists of billions of records in the form of data collections called "linked data". With the ever-increasing linked big data on the Web, it is imperative to process this data with powerful and scalable techniques in distributed processing environments such as MapReduce. There are several distributed RDF processing systems, including SemaGrow, FedX, SPLENDID, PigSPARQL, SHARD, SPARQLGX, that are developed over the years. However, there is a need for computational and qualitative comparison of the differences and similarities among these systems. In this paper, we extend a previous comparative analysis to a diverse study with respect to qualitative and quantitative analysis views, through an experimental approach for these distributed RDF systems. We examine each of the selected RDF query systems with respect to the implementation setup, system architecture, underlying framework, and data storage. We use two widely used RDF benchmark datasets, FedBench and LUBM. Furthermore, we evaluate and examine their performances in terms of query execution time, thus, analyzing how those different types of large-scale distributed query engines, support long-running queries over federated data sources and the query processing times for different queries. The results of the experiments in this study show that SemaGrow distributed system performs more efficiently compared to FedX and Splendid, even though in smaller queries the former performs slower.

Keywords: Linked Data · Big Data · Semantic Web · Resource Description Framework (RDF) · SPARQL Protocol and RDF Query Language · Distributed RDF Query Processing · Triple Pattern (TP)

1 Introduction

Semantic Web is the next-generation web that provides a standard method for structuring data with a set of protocols for querying, processing, and extracting meaningful

H. Han and E. Baker (Eds.): SDSC 2022, CCIS 1725, pp. 158–170, 2022.
https://doi.org/10.1007/978-3-031-23387-6_11

information using reasoning and description logic based on a specific data model. It is not an independent web from the current web, but rather an extension of it with varying capabilities [1]. Semantic Web framework provides a number of standard protocols to define and structure data, and they are RDF,[1] RDFS,[2] and OWL.[3] Resource Description Framework (RDF) is a World Wide Web Consortium (W3C) standard for defining data nodes and the relationships between them in a triple form of *subject-predicate-object* entities that codifies a statement for semantic web data. It is also an exchange model that represents data as a graph, as illustrated in Fig. 1. RDF graph nodes and edges have no internal structure, rather a *subject* is a resource or node in the graph representing a real-world object or concept. A *predicate* represents a relationship edge between two nodes, and an *object* is the other node in this relationship, it can be another resource or a literal value. Resource Description Framework Schema (RDFS) is a general-purpose language for representing RDF data vocabularies and semantic extensions. It describes related resource groups with the relationships among them, and the characteristics of other resources, as well as the domain links with the range of properties [1]. However, there exist expression limitations with RDF Schema. To solve this, W3C extended the previous RDF Schema framework to a more three-layered architecture standard regarded as Web Ontology Language (OWL) [2, 3].

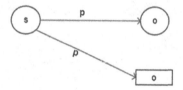

Fig. 1. RDF graph.

Moreover, W3C sets the syntax and semantics of a standard query language for RDF specification called SPARQL,[4] short for (*SPARQL Protocol And RDF Query Language*). It is based on graph pattern specifications that provide conjunction, disjunction, aggregation, negation, and sub-query functionalities to mention a few. Over the years, many SPARQL Endpoints[5] are provided to enable a point of reference on an HTTP network that is capable of receiving and processing SPARQL query requests against semantic web data stores. A SPARQL endpoint is identified by a URL commonly referred to as a SPARQL Endpoint URL. This endpoint enables access to RDF statements or sentences (data) by, for example, federated SPARQL queries remotely.

The common query standard of SPARQL can be described as follows:

QUERY : = {*select* RD *where* GP}

---

[1] RDF: https://www.w3.org/TR/rdf-primer/.
[2] RDFS: https://www.w3.org/TR/rdf-schema/.
[3] OWL: https://www.w3.org/TR/owl-features/.
[4] SPARQL: https://www.w3.org/TR/sparql11-query/.
[5] DBpedia SPARQL endpoint: http://dbpedia.org/sparql.

where GP are *triple patterns* and RD is the *result description*.

Furthermore, Table 1 illustrates examples of the standard form of a triple in an RDF database that represents a relationship, indicated by the triple pattern *predicate*, between the *subject* and the *object*, a directed relationship from subject to object. In addition to that, Fig. 2 shows the graph representation of the same RDF data, in which subjects and objects are the nodes in the graph, and each edge connects two nodes as a predicate. On the other hand, an example of a SPARQL query of finding an experienced person in 'Hims' (Health Information Management System) is shown in the Listing 1.

RDF is the common data model for almost all semantic data sources and online public datasets such as online bioinformatics resources. Many open-source RDF linked data can be found over the web, for example in Linked Open Data Cloud[6] project, listing datasets such as DBPedia,[7] Uniprot,[8] Bio2RDF,[9] OrthoDB,[10] Rhea Reaction Data,[11] and many more.

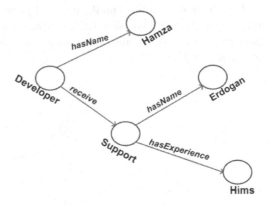

Fig. 2. RDF graph example.

Table 1. Sample RDF statements.

ID	Subject	Predicate	Object
1	*<Developer>*	*<hasName>*	*"Hamza"*
2	*<Developer>*	*<receive>*	*<Support>*
3	*<Support>*	*<hasName>*	*"Erdogan"*
4	*<Support>*	*<hasExperience>*	*<Hims>*

---

[6] https://lod-cloud.net/.

[7] https://wiki.dbpedia.org.

[8] https://www.uniprot.org/.

[9] https://bio2rdf.org/.

[10] https://www.orthodb.org/.

[11] https://www.rhea-db.org/.

```
PREFIX:<http://www.semanticweb.org/hamza.mohammed/2020/3/Company#>
PREFIX rdf: <http://www.w3.org/1999/02/22-rdf-syntax-ns#>
PREFIX rdfs: <http://www.w3.org/2000/01/rdf-schema#>
PREFIX owl: <http://www.w3.ord/2002/01/owl#>
SELECT ?p ?name
WHERE {
    ?p <hasName> ?name .
    ?p <hasExperience> <Hims>
}
```

Listing 1. Sample SPARQL query

Traditionally, RDF stores are designed as centralized systems, like in RDF-3X [4], and SW-Store [5]. However, recent rapid increases in the volume of RDF datasets and the complexity of query operations require most RDF systems to perform with low latency and high throughput, especially regarding concurrent queries. Therefore, distributed data stores are designed to improve efficiency for querying RDF data, and at the same time achieve high throughput. For example, SHARD [6], DREAM [7], and TriAD [8] implemented their systems in a distributed architectural design. More recently, big data interest keeps gaining attraction due to the high volume of RDF data and an exponential increase in web content and other digital platforms. In response to this, other distributed systems were introduced, such as SPARQLGX [9], SemaGrow [10], and Wukong [11]. In this study, we selected some of these popular distributed systems and highlighted their underlying design features and their responsiveness to SPARQL queries on RDF datasets.

We present the related studies in Sect. 1 We describe our methodology in Sect. 2 and present the evaluation results in Sect. 3, and finally conclude in Sect. 4.

2 Related Work

Many studies have been conducted related to distributed RDF systems, such as an overview of centralized and distributed RDF systems along with their respective query processing technique utilized on linked data [12]. Ghaleb et al. [13] also provided a study for representing RDF as a graph or hyper-graph data model. They provided two ways to convert relational database into RDF model while addressing some database features like function dependency, information preservation, query preservation, integrity, normalization, re-usability, and semantics preservation. Another survey [14] provided techniques for using relational and NoSQL databases to store RDF data. Furthermore, Azzam et al. [15] proposed a new system called FLINKer, to provide an RDF query and processing via Flink[12] native streaming and processing framework. They provide ways to integrate and extend Flink's graph processor architecture with centered distributed graph processing to achieve SPARQL queries over RDF data. They also highlighted some challenges and potential optimization problems while adopting their methodologies.

---

[12] https://flink.apache.org/.

In [16] Ozsu presents a survey of state-of-the-art parallel and distributed RDF data management systems designed and deployable in the cloud that consist of complex architectures with different data storage, query processing, and reasoning capabilities. It also emphasizes the need for future distributed and parallel RDF systems that have parallel RDF data management capabilities and some optimization techniques in query decomposition and join order in order to keep up with the other distributed database systems [17]. Lastly, [18] provides a general overview of indexing methods on linked data and presents new indexing techniques with a prototype implementation that possesses efficient and scalable data distribution methods.

Nevertheless, none of the previously stated studies provide an extensive overview of the distinctive comparative features of these systems on the performance and the underlying architectural design. In this paper, we provide an explicit study based on experimental analysis and architectural assessment with respect to different RDF queries. We used large benchmark datasets consisting of millions of triples to evaluate the systems with various queries.

3   Method

Distributed RDF systems partition RDF graph data across multiple nodes as sub-graphs. Then individual SPARQL queries are broken down into sub-queries and evaluated in parallel on sub-graphs. A continuous exchange of communication between nodes throughout the distributed query evaluation takes place in this system.

3.1   Data Storage Format, Query Planning, Evaluation, and Optimization

Processing RDF data requires a certain suitable data storage format to perform operations on them, especially when it comes to distributed query processing. The systems we use in this study use different architectures. Here we review each briefly:

FedX[13]**:** This is a widely known, effective distributed query processing system over linked data [19]. The system architecture implementation of FedX combines previous optimization techniques based on Sesame[14] in a distributed setting. The system integrated Sesame's Storage and Inference Layer (SAIL) to allow the incorporation of standard and customized RDF repositories mechanism. It also adopts a heuristic approach in order to optimize the query plan without relying on metadata. The system uses suboptimal algorithms in order to avoid having exponential query processing times. Nevertheless, a vast amount of storage and modeling overhead problem exists in this system. This's due to caching query results. They proposed the usage of Semantic Web Rule Language (SWRL), which reduces the modeling overhead. This allows state compatibility on the class level [20]. However, the efficient query execution engine makes little overhead, affecting the speed of execution drastically. Thus, FedX offers fast distributed execution of RDF query processing and evaluation.

---

[13] https://github.com/VeritasOS/fedx.
[14] https://www.w3.org/2001/sw/wiki/Sesame.

SemaGrow[15,16]**:** SemaGrow [10] aims to optimize query processing time due to high computational complexity and complex query planning of distributed queries. The system architecture adopts an asynchronous stream processing mechanism for the query execution engine and a non-blocking communication mechanism between the query optimizer and the query planner. These different features and the novel design mechanism in the system result in robust and efficient querying of distributed RDF data compared to the other systems. Nonetheless, high overhead exits in SemaGrow as per the absence of metadata, even though the system achieves the best optimal query plan compared to FedX and Splendid [10]. Furthermore, there are also no client privileges for data storage organization among the nodes.

Implemented design in SemaGrow system is similar to FedX and SPLENDID, for example, ignoring the join operation in the optimization plan. However, this system adopted a naive approach during query operation compared to those of the other two systems mentioned. First, source selection is made on the query to shrink the query execution scope, and then it applies the query operation on the data at the next step. The system adopts a generic dynamic algorithm for join ordering in queries with join operations. It selects the query part with a minimum cost of n number of triple patterns and builds up in a bottom-up way. Despite the algorithm tracking minimal space possible, the enumeration of all possible plans is exponential in this system.

SPLENDID[17]**:** This is another Federated SPARQL query system that optimizes query execution on RDF-linked data [21]. The SPLENDID system makes use of metadata of the RDF data to optimize the query execution time, and it does not leverage non-blocking and asynchronous stream processing techniques to achieve flexibility and robustness in the query execution compared to the Semagrow system. However, SPLENDID relies on VoID metadata of federated data sources to perform a query. It provides the resources and tools required to retrieve the metadata that it needs automatically, nonetheless using those tools or resources requires access to data dumps. The system uses open semantic web standards, such as VoID and SPARQL endpoints, which allow for efficient integration or incorporation of various distributed and related RDF data sources. Although both hash join and bind join considerably reduce the processing time of complex query types, this system does not yet optimize the actual query execution.

Wukong[18]**:** **Wukong** [11, 22] is a distributed in-memory RDF store that leverages RDMA-based graph exploration to support fast and concurrent RDF queries. Wukong term originated from a Sun Wukong character known for his breakneck speed (21,675 km in one somersault) and the ability to fork himself to do massive multitasking. This system stores the RDF data in a directed graph in the same context as the sketch illustrated in Fig. 1, using two indexing techniques. One is predicate indexing, in which the indexing vertex serves as an opposite vertex to the subjects or objects, And the second type of indexing is to group a set of subjects. The storage type, indexing techniques, and allocating strategy features differentiate this system from other graph-based systems.

---

[15] http://semagrow.github.io/deployment-instructions/.

[16] https://github.com/semagrow/semagrow.

[17] https://github.com/goerlitz/rdffederator.

[18] http://ipads.se.sjtu.edu.cn/projects/wukong.

The designers of this system claim to achieve orders-of-magnitude lower latency and high throughput than prior state-of-the-art systems.

Wukong+G[19]: This is an extension of the original Wukong [11] distributed RDF store with a distributed GPU support capability [22, 23]. This is also a graph-based distributed RDF query processing system that efficiently exploits the hybrid parallelism of CPU and GPU. Wukong+G is a fast and concurrent system with three key design features. First, Wukong+G utilizes GPU to tame random memory accesses in graph exploration by efficiently mapping data between CPU and GPU for latency hiding, including a set of techniques such as query-aware pre-fetching, pattern-aware pipeline, and fine-grained swapping. Second, Wukong+G scales up by introducing a GPU-friendly RDF store to support RDF graphs exceeding GPU memory size, using techniques like predicate-based grouping, pairwise caching and look-ahead replacing to narrow the gap between host and device memory scale. And lastly, it scales out by adopting a communication layer in order to decouple the transfer process for query metadata and intermediate results and leverages both native and GPU Direct RDMA to enable efficient communication on a CPU/GPU cluster.

TriAD [8]: is another distributed RDF system that scales. The architecture of this system applies traditional RDF graph summary to the RDF query and then combines horizontal partitioning of the RDF data so that the triples transform into a grid distributed index structure. Triad joins operations are executed in a multithreaded, distributed form. It is enabled by the message passing protocol in asynchronous ways to run multiple join operators completely in parallel, along with a request schedule. The designers believe that the architecture of this system offers a unique approach to join-ahead pruning in a distributed environment, as the more conventional or traditional method of sideways passing information does not allow distributed joins to be performed in an asynchronous manner.

SPARQLGX[20]: SPARLQLGX [9] is an Apache Spark[21] implementation based on vertical partitioning methodology for providing SPARQL query processing on RDF data similar to the to the recently introduced method by Damien et al. [24]. The rampant increase in RDF data resulted in the implementation of the so-called SPARQLGX - a distributed RDF data store. They designed this system in such a way it directly translates SPARQL queries into an executable Apache Spark plane code that composes evaluation mechanisms in terms of storage method and statistical data. With the storage type, the system utilizes the HDFS[22] storage file system so that a triple file (subject, predicate, object) is stored in a file named after the predicate, and it contains only similar subject and objects entries. This storage mechanism results in less space usage in a file system, given that data is compressed as the predicate elements column is eliminated. This also leads to a shorter response time and, more importantly, a fast conversion rate, given that only one dataset traversal is required for a particular query on RDF data. Moreover,

---

[19] https://github.com/WukongGPU/WukongGPU.

[20] https://github.com/tyrex-team/sparqlgx.

[21] https://spark.apache.org/.

[22] https://hadoop.apache.org/docs/r1.2.1/hdfs_design.html.

SPARQLGX translates triple patterns directly to Apache Spark syntax primitive definitions, after which common variables will be joined together in the compilation process
with a list of filters. The query order of SPARQL affects query performance, especially
when dealing with complex queries on large datasets. With the Spark evaluation process,
first triple patterns are evaluated and then it is followed by subset joins with respect to
common variables. This in turn resulted in shorter evaluation times because communication between the worker nodes is faster. It also optimizes join operation using minimal
selection criteria. This system concept can be denoted as:

Given RDF dataset D with T number of triples, and a place in the RDF $k \in \{subject,$
$predict, object\}$, with tuple $\{subject, predict, object\}$ over the D dataset can describe as
follows:

$$sel_D(x,y,z) = min(sel_D^{subject}(x), sel_D^{predict}(y), sel_D^{object}(z)).$$

The formula above defines the selectivity in D of an element e located at k $(sel_D^k(e))$
by SPARQLGX as:

– the occurrence number of e as k in D if e is a constant
– T if e is a variable

Likewise, the formula define the selectivity of triple pattern $TP(x,y,z)$ over an RDF
dataset (D). Thus, calculates statistics about the number of data records (datasets) across
all distinct *subjects, predicates, and objects*. This is implemented in a compile-time
module that sorts triple patterns (tps) in ascending order of selectivity before compilation
(translated).

The system also has a direct evaluator based on the same SPARQL translation mechanism, called SDE, for circumstances where pre-processing time is as important as query
evaluation time. Furthermore, the execution order of the query can also have an impact on
performance. As described previously, the evaluation process (using Spark) first evaluates triple patterns (tps) and then joins these subsets according to their common variables.
Thus, minimizing the intermediate set sizes involved in the join process reduces evaluation time (since communication between workers is then faster). Thereby, statistics on
data and information on intermediate results sizes provide useful information that they
exploit for optimization purposes.

4 Evaluation

We examine and evaluate the performance of different distributed RDF systems. These include SemaGrow, stacked together with FedX, SPLENDID (stacked version) using the experimental setup[23] provided, and then SPARQLGX, Triad, Wukong, and its subsequent version Wukong+G.

4.1 System Setup

The experiments are carried out on Ubuntu Linux 64-bit 20.04.4 LTS running on an Intel Core i7-4712 processor. In order to utilize memory efficiently and deal with the complexity of the configuration of system tools, libraries, and settings, we used the Docker container tool.[24]

4.2 Dataset

In the experiments, thanks to the stack version of Federated Semagrow [10] RDF query processing system implementation, comprising both SPLENDID and FedX implementations composed together, we did not need too much configuration. With that, we used directly the FedBench benchmark of millions of triples to evaluate the three federated systems provided in the repository.[25] FedBench consists of different forms of queries, from simple to complex queries. The queries are labeled based on star type, chain-star, and complex, in accordance with the query formulation, and with respect to complexity and use of UNION and JOIN operations. The datasets and queries are available in the Fed-Bench repository.[26] The datasets and the corresponding queries listed in Table 3 consist of *Cross-Domain (CD)* and *Life Sciences (LS)* datasets. *CrossDomain* dataset comprises a subset of DBpedia,[27] NY Times, LinkedMDB, Jamendo, GeoNames, and SW Dog-Food, while *Life Science* dataset includes a subset of DBpedia, KEGG, Drugbank, and ChEBI benchmarks. Some queries are cross-dataset queries on the above benchmark datasets, and the construct structure queries are of various range types. Some are simple forms, and the others are chain queries with complex query plans as described earlier. Moreover, we perform a second experiment for the SPARQLGX system separately. In the experiment, we use LUBM [22] benchmark of almost 1.36 billion triples of universities (Table 2).

---

[23] https://github.com/semagrow/kobe.
[24] https://www.docker.com/.
[25] https://code.google.com/archive/p/fbench/.
[26] https://code.google.com/archive/p/fbench/downloads.
[27] http://ontotext.com/factforge/.

Table 2. Dataset characteristics.

	Collection	Dataset	Triples (millions)	File size on HDFS (GB)
Lehigh University Benchmark	LUBM	*LUBM*	109	46.8 GB
FedBench	Cross Domain (CD)	*DBpedia subset*	43.6	–
		NY Times	0.335	–
		LinkedMDB	6.15	–
		Jamendo	1.05	–
		GeoNames	108	–
		SW Dog Food	0.104	–
	Life Science (LS)	*DBpedia subset*	43.6	–
		KEGG	1.09	–
		Drugbank	0.767	–
		ChEBI	7.33	–

4.3 Results

The results in Table 3 show the query processing times for FedBench queries CD1-7 and LS1-7. The results show that FedX and SemaGrow always perform better than SPLENDID in terms of processing times of queries. We also conclude that among the three RDF systems, Semagrow is more efficient in query processing times when compared to the others. There are some huge differences in processing some queries, i.e., for query LS6, FedX processes in 22321 ms whereas SemaGrow processes the same query in just 98 ms.

On the other hand, in Table 4 we present query processing times for SPARQLGX, Wukong, Wukong+G, and TriAD systems on LUBM dataset. The results show the total processing time of the queries for the system. The direct query assessment and evaluation of the selected large queries of LUBM as the statistical characteristic with respect to their respective complexities. Q1-Q3 and Q7 are large queries that involve a large subset of the RDF graph, while Q4-Q6 consists of small queries with simple queries on a small subset graph. According to the results, the queries Q1, Q2, and Q7 show speedup between 1.51–4.3 between Sparqlgx versus its adjacent system Wukong+G in terms of query performance. Likewise, the geometric values in the table indicate that Sparqlgx outperforms the other systems. The result shows a query latency reduction of 1.27 compared to the Wukong+G system with regards to both respective geometric mean values.

Table 3. FedBench Benchmark: Query processing time (msec).

Query	Query processing system		
	SemaGrow	FedX	SPLENDID
CD1	**20**	28	75
CD2	**13**	19	43
CD3	**45**	57	73
CD4	49	**38**	51
CD5	96	**39**	46
CD6	**402**	429	10735
CD7	**503**	523	2453
LS1	72	**42**	54
LS2	43	**34**	193
LS3	**3042**	3952	44233
LS4	89	**31**	74
LS5	**2010**	98,728	297484
LS6	**98**	22,321	18634
LS7	**1203**	20,135l	19321
Geometric Mean	147.05	310.74	810.88

Table 4. Lehigh University Benchmark: Query processing time (msec).

Dataset	Query processing system				
	Query	Sparqlgx	Wukong+G	Wukong	Triad
LUBM	Q1	**109**	165	992	851
	Q2	100	**31**	138	211
	Q3	–	**63**	340	424
	Q7	**23**	100	828	2194
	Q9	18	–	–	–
	Q13	21	–	–	–
	Q14	17	–	–	–
Geometric Mean (Q1, Q2 and Q7)		**63.05**	79.97	653.60	730.19

5   Conclusion

In this paper, we presented an overview of the selected RDF distributed systems. We
introduced each system briefly and then discussed the similarities and differences. We

highlight some optimization techniques in use for join operations in these systems as well. Almost all the systems experimented with in this paper are open-source projects, therefore they can be accessed through the references for future work.

Despite OWL, Linked Data, and triples-based RDF standard models having been around for some time in the research community and are considered to be successful in many domains, especially in healthcare, there is still a lack of expanded analysis of these systems. Therefore, we plan to expand this research to further research and perform an advanced comparative analysis of these systems, including federated SPARQL engines [25–29]. In particular, more recent federated engines, such as CostFed and Odyssey. Similarly, a recent survey [22, 30] on SPARQL query processing engines highlighted new recent distributed systems, e.g. SANSA, DISE, WISE, Triag, gSmart, which we plan to include and evaluate as well. There also exist more recent benchmarks for federated SPARQL query processing, e.g. LargeRDFBench. Similarly, for the triple store evaluation, we plan to use a feasible real-world SPARQL benchmark along with WatDiv [23] to complement both real-world and synthetic state-of-the-art benchmarks.

References

1. Sen, S., Malta, M.C., Dutta, B., Dutta, A.: State-of-the-art approaches for meta-knowledge assertion in the web of data. IETE Tech. Rev. **38**(6), 672–709 (2021)
2. Yumusak, S., Kamilaris, A., Dogdu, E., Kodaz, H., Uysal, E., Aras, R.E.: A discovery and analysis engine for semantic web. In: Companion Proceedings of the Web Conference, pp. 1497–1505 (2018)
3. Jevsikova, T., Berniukevičius, A., Kurilovas, E.: Application of resource description framework to personalise learning: systematic review and methodology. Inform. Educ. **16**(1), 61–82 (2017)
4. Ali, W., Saleem, M., Yao, B., Hogan, A., Ngomo, A.C.: Storage, indexing, query processing, and benchmarking in centralized and distributed RDF engines: a survey. (2020)
5. Ben Mahria, B., Chaker, I., Zahi, A.: An empirical study on the evaluation of the RDF storage systems. J. Big Data. **8**(1), 1–20 (2021)
6. Hassan, M., Bansal, S.K.: S3QLRDF: property table partitioning scheme for distributed SPARQL querying of large-scale RDF data. In: 2020 IEEE International Conference on Smart Data Services (SMDS) pp. 133–140 (2020)
7. Hammoud, M., Rabbou, D.A., Nouri, R., Beheshti, S.M., Sakr, S.: DREAM: distributed RDF engine with adaptive query planner and minimal communication. Proc. VLDB Endow. **8**(6), 654–65 (2015)
8. Wylot, M., Hauswirth, M., Cudré-Mauroux, P., Sakr, S.: RDF data storage and query processing schemes: a survey. ACM Comput. Surv. (CSUR). **51**(4), 1–36 (2018)
9. Graux, D., Jachiet, L., Genevès, P., Layaïda, N.: Sparqlgx: efficient distributed evaluation of sparql with apache spark. In: International Semantic Web Conference, Springer, Cham, pp. 80–87 (2016)
10. Charalambidis, A., Troumpoukis, A., Konstantopoulos, S.: SemaGrow: optimizing federated SPARQL queries. In: Proceedings of the 11th International Conference on Semantic Systems, pp. 121–128 (2015)
11. Shi, J., Yao, Y., Chen, R., Chen, H., Li, F.: Fast and concurrent {RDF} queries with {RDMA-Based} distributed graph exploration. In: 12th USENIX Symposium on Operating Systems Design and Implementation (OSDI 16), pp. 317–332 (2016)

12. Özsu, M.T.: A survey of RDF data management systems. Front. Comput. Sci. **10**(3), 418–32 (2016)
13. Ghaleb, F.F., Taha, A.A., Hazman, M., Abd ElLatif, M.M., Abbass, M.: A comparative study on representing RDF as graph and hypergraph data model (2019)
14. Ma, Z., Capretz, M.A., Yan, L.: Storing massive Resource Description Framework (RDF) data: a survey. Knowl. Eng. Rev. **31**(4), 391–413 (2016)
15. Azzam, A., Kirrane, S., Polleres, A.: Towards making distributed rdf processing flinker. In: 2018 4th International Conference on Big Data Innovations and Applications (Innovate-Data) IEEE, pp. 9–16 (2018)
16. Li, R., Mo, T., Yang, J., Jiang, S., Li, T., Liu, Y.: Ontologies-based domain knowledge modeling and heterogeneous sensor data integration for bridge health monitoring systems. IEEE Trans. Ind. Inform. **17**(1), 321–32 (2020)
17. Valduriez, P., Jiménez-Peris, R., Özsu, M.T.: Distributed database systems: the case for newSQL. In: Transactions on Large-Scale Data-and Knowledge-Centered Systems XLVIII, pp. 1–15. Springer, Berlin, Heidelberg (2021)
18. Svoboda, M., Mlýnková, I.: Linked data indexing methods: a survey. In OTM Confederated International Conferences" On the Move to Meaningful Internet Systems", Springer, Berlin, Heidelberg, pp. 474–483 (2011)
19. Schwarte, A., Haase, P., Hose, K., Schenkel, R., Schmidt, M.: FedX: a federation layer for distributed query processing on linked open data. In: Extended Semantic Web Conference, Springer, Berlin, Heidelberg, pp. 481–486 (2011)
20. Horrocks, I., Patel-Schneider, P.F., Boley, H., Tabet, S., Grosof, B., Dean, M.: SWRL: a semantic web rule language combining OWL and RuleML. W3C Memb. Submiss. **21**(79), 1–31 (2004)
21. Qudus, U., Saleem, M., Ngonga Ngomo, A.C., Lee, Y.K.: An empirical evaluation of cost-based federated SPARQL query processing engines. Semant. Web. **12**(6), 843–868 (2021)
22. Ali, W., Saleem, M., Yao, B., Hogan, A., Ngomo, A.C.: A survey of RDF stores & SPARQL engines for querying knowledge graphs. VLDB J. **13**, 1–26 (2021)
23. Wang, S., Lou, C., Chen, R., Chen, H.: Fast and concurrent {RDF} queries using {RDMA-assisted} {GPU} graph exploration. In: 2018 USENIX Annual Technical Conference (USENIX ATC 18), pp. 651–664 (2018)
24. Naacke, H., Curé, O.: On distributed SPARQL query processing using triangles of RDF triples. Open J. Semant. Web (OJSW). **7**(1), 17–32 (2020).
25. Cheng, S., Hartig, O.: FedQPL: a language for logical query plans over heterogeneous federations of RDF data sources. In: Proceedings of the 22nd International Conference on Information Integration and Web-based Applications & Services, pp. 436–445 (2020)
26. Heling, L., Acosta, M.: Cost-and robustness-based query optimization for linked data fragments. In: International Semantic Web Conference, Springer, Cham, pp. 238–257 (2020)
27. Potoniec, J.: Learning OWL 2 property characteristics as an explanation for an RNN. Bull. Pol. Acad. Sci.: Tech. Sci. (6) (2020)
28. Haller, A., Fernández, J.D., Kamdar, M.R., Polleres, A.: What are links in linked open data? A characterization and evaluation of links between knowledge graphs on the web. J. Data Inf. Qual. (JDIQ). **12**(2), 1–34 (2020)
29. Polleres, A., Kamdar, M.R., Fernández, J.D., Tudorache, T., Musen, M.A.: A more decentralized vision for linked data. Semant. Web **11**(1), 101–13 (2020)
30. Ji, S., Pan, S., Cambria, E., Marttinen, P., Philip, S.Y.: A survey on knowledge graphs: representation, acquisition, and applications. IEEE Trans. Neural Netw. Learn. Syst. **33**(2), 494–514 (2021)

Normal Equilibrium Fluctuations from Chaotic Trajectories: Coupled Logistic Maps

Kyle Taljan[1] and J. S. Olafsen[2]($\boxtimes$)

[1] Case Western Reserve University, Yost Hall 2049 Martin Luther King Jr. Drive, OH 44106-7058 Cleveland, USA
[2] Department of Physics, Baylor University, TX 76798 Waco, USA
Jeffrey_Olafsen@baylor.edu

Abstract. We report results of a numerical algorithm to examine coupling of two logistic maps where the mixing is chosen to maintain the stability of one map at the loss of the other. The long-term behavior of the coupling is found to contain windows in which the mixing results in Gaussian fluctuations about a fixed point for the stabilized map. This deterministic behavior is the result of the destabilized map simultaneously being driven into a chaotic regime and not noise. The results are applicable to both chaotic encryption of data and recapturing equilibrium behavior in a non-equilibrium system.

Keywords: Encryption · Nonlinear dynamics · Logistic maps

1 Introduction

Chaotic maps can be thought of as simple systems that demonstrate non-equilibrium behavior: deterministic systems that nonetheless take an essentially infinite amount of time to repeat themselves. Chaotic behavior can thus be classified as a certain type of steady-state dynamics that is the result of a small number of inputs. This picture is particularly beneficial in using the logistic map

$$Z_n + 1 = \mu z_n(1 - z_n) \tag{1}$$

as a simple system that at low values of the parameter μ demonstrates equilibrium behavior in the long-term mapping to a single point and demonstrates a period doubling path to chaos as μ is increased [1]. Studying non-equilibrium systems contributes to a better physical description of their poorly understood thermostatistics. While it has been demonstrated that a proper selection of coarse graining [2] or particular control of the type of non-equilibrium balance between energy injection and dissipation [3] can result in a recapturing of equilibrium-like behavior and Maxwell-Boltzmann statistical fluctuations, a fundamental picture of the laws of non-equilibrium thermodynamics remains elusive. Chaotic maps, which themselves may be thought of as a subset of non-equilibrium systems [4], can also potentially aid in the encryption of information

© The Author(s), under exclusive license to Springer Nature Switzerland AG 2022
H. Han and E. Baker (Eds.): SDSC 2022, CCIS 1725, pp. 171–179, 2022.
https://doi.org/10.1007/978-3-031-23387-6_12

for security purposes [5]. A set of coupled chaotic maps demonstrating an underlying Gaussian dynamic would also be beneficial for use as a deterministic manner in which to hide information in a signal that appears random [6].

Motivated by recent observations of a granular dimer on a vertically shaken plate [7], a set of coupled logistic maps were proposed as a potentially simple model of a two-component system with both a stable and unstable dynamic. This choice is because a single particle on an oscillating plate demonstrates a period doubling path to chaotic behavior. The shaken dimer demonstrates a breaking of symmetry where one sphere of the dimer appears to remain nearly stable with the shaken plate while the other chatters with a phase coherence that may demonstrate chaotic instability [8]. To model this dynamic, we proposed a specific pair of coupled logistic maps:

$$x_{n+1} = \mu_x x_n (1 - x_n) + \varepsilon x_n y_n \tag{2}$$

$$y_n + 1 = \mu y_n (1 - y_n) - \varepsilon x_n y_n \tag{3}$$

where the $\pm \varepsilon x_n y_n$ term provides the mixing between the two maps, x and y, via the small positive parameter, ε. Since the two uncoupled maps are independent and exist in a regime of [0,1], the positive mixing term effectively increases the value of μ_x in Eq. 2 and destabilizes the x map by driving it further along the period doubling path to chaos while at the same time effectively decreases μ_y in Eq. 3 and stabilizes the y map for small values of ε. This is an example of a "master-slave" system [9, 10].

While the results will demonstrate a more immediate application to the encryption of data, initially this mixing was developed as a potential model of the effective cross term responsible for the anisotropic behavior observed in a dimer on a vertically shaken plate [11]. Because of this motivation, the μ_y values studied here were limited to a range over which the logistic map demonstrates a single stable fixed point, and μ_x was allowed to vary over a wider range of values. However, because the logistic map is only properly defined on the range of [0, 1], the mixing parameter ε was purposely kept small in this study and limited to values in the range of [0, 0.075]. In this regime, the x map is destabilized while the net effect on the y map is to maintain the stability in a window about the fixed point. These choices also avoided errors with the double-precision calculation of the trajectories [12].

For $\varepsilon = 0$, the two maps are uncoupled and the typical logistic map behavior is of course recovered for each of $x(\varepsilon = 0) = y(\varepsilon = 0) = z$ as shown in Fig. 1. The squares in the range of $2.8 \leq \mu \leq 3.05$ denote the values for μ_y used in this study and the blue region centered about $\mu \sim 3.8$ the values of interest for μ_x that will be discussed later along with specific values of μ_x and μ_y indicated by the arrows. For comparison, when the mixing is turned on by allowing $\varepsilon > 0$, the behavior is quite different as demonstrated in Fig. 2(a) and (b) for the long-term trajectories of x and y, respectively, as a function of μ_x. (For the rest of this letter, the prescription was to select a value of μ_y that kept y at or near a stable fixed point, as shown as boxes in Fig. 1 for the uncoupled logistic map, and then to examine the coupled equations as a function of the parameter μ_x.) In Fig. 2, the value of μ_y is 2.95 and ε is 0.05. Careful examination of Fig. 2(a) for the coupled x map compared to the uncoupled z map of Fig. 1 confirms the general behavior of ε for the x map is to effectively *increase* the value of μ_X. Note that the bifurcation to period 2 behavior that occurs in the uncoupled map for $\mu = 3.0$ in Fig. 1 occurs for the coupled x map in Fig. 2(a) at a slightly lower value of $\mu_X \sim 2.975$. In nearly every regard, the bifurcation diagram of x is simply the same as the uncoupled logistic map shifted slightly lower in μ_X (made more unstable) by the addition of the mixing term. (As an additional guide to the eye, the vertical arrow to the right side of Fig. 1 is at the same μ value as the vertical arrow in Fig. 2(a).) In Fig. 2(a) and 2(b), the long-term behavior of each trajectory at each μ_X is demonstrated by iterating the maps for 30,000 steps and plotting the last 20,000 iterations of each trajectory.

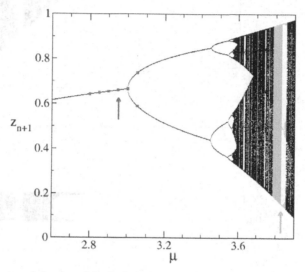

Fig. 1. The long-term behavior of the logistic map as a function of the parameter μ. The boxes on the left denote the μ_Y values used for the coupled y-map, while the shaded region on the right denotes the parameter space of interest for μ_X of the coupled x-map.

Far more interesting is the effect the mixing term has on the behavior of the y map. Even though μ_Y for the y map is held fixed at 2.95, and the lowest order effect of the $-\varepsilon\, x_n\, y_n$ term is to maintain that stability by effectively *reducing* μ_Y, the coupling adds fluctuations (from the x map) about what would be a stable fixed point without the coupling. In the periodic regions, these fluctuations simply behave as a self-similar version of the logistic map superimposed upon the stable fixed point of the uncoupled y map for $\mu_Y = 2.95$. One will also note that while the value of the fixed point moves for μ values in the range between 2.6 and 2.9 in Figs. 1 and 2 (a), the fixed point in the y map is stationary in Fig. 2(b). Yet, in the chaotic regions of the x map, the mixing term produces fluctuations that are different from the uncoupled logistic map in that they remain in a small window (of approximate size $\pm\, 0.05$) about the stable fixed point for the uncoupled y map. The width of this window is determined by the value of ε.

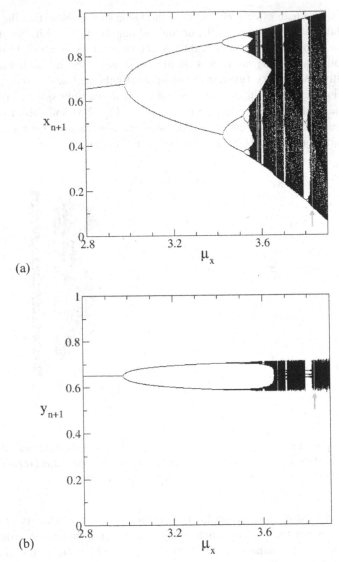

x_{n+1}

μ_x

(a)

y_{n+1}

μ_x

(b)

Fig. 2. The long-term behavior for the coupled maps (a) x_n and (b) y_n as a function of μ_X with $\mu_Y = 2.95$ and $\varepsilon = 0.050$. The arrows denote the value of μ_X of interest discussed in Figs. 3 and 4.

2 Discussion

To characterize the fluctuations of the coupled y map about the fixed point from its uncoupled counterpart, one can examine the flatness of the distribution. The flatness of a set of values, v, is defined as the ratio of the fourth moment of the fluctuations from the mean of v to the square of the variance, or second moment, of the fluctuations of v:

$$F = <v^4> / <v^2>^2 \qquad (4)$$

where a flatness F = 3 is the result for a Gaussian (normal) distribution [13]. Figure 3 is a plot of the flatness as determined by the second and fourth moments of the last 20,000 iterations of the long-term behavior of the coupled maps. (Note: The value of F – 3, demonstrating the deviation from Gaussian or the kurtosis, is what is plotted.) The lighter (blue) line is the result for the coupled x map, which is simply a shifted version of the uncoupled logistic map, while the darker (black) line is the result for the coupled y map. For both the x and y maps, in regions where the maps are single valued, the fourth and second moments are equal, resulting in a flatness of unity, so F – 3 = –2. The relative behavior of the flatness for the x and y maps in the range of 3.4 < μ_X < 3.7 also underscores that the y map is more stable than the x map by introduction of the mixing term.

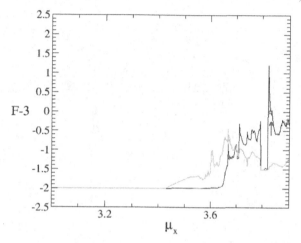

Fig. 3. The flatness of the distribution of x_n (blue) and y_n (black) points from the long-term behavior of the maps. The box and arrow denotes a situation where the y_n points form a Gaussian distribution.

Yet, in the regime of μ_X where the coupled y map exhibits period doubling, the flatness of the y map is generally closer to Gaussian than that for either the coupled x map or the uncoupled logistic map (not shown). Indeed, at just above μ_X = 3.8, the plot of F – 3 of the fluctuations about the mean value for the y map rises above zero, before passing back through zero and becoming negative. The point where this occurs is denoted by the box and arrow in Fig. 3 as well as the vertical arrow in Fig. 2(b). This is a dynamic result of the mixing that occurs for a particular value of the parameters and is not due to additive noise [14] or computational imprecision [12].

At μ_X = 3.832, the flatness of the fluctuations for the y map about its mean indicates that the tails of the distribution are nearly Gaussian. Further work was done to investigate whether this was a simple happenstance or if there were other parameter values for which the deterministic fluctuations about the mean value of the y map were Gaussian. Table 1 lists several values of the parameters μ_X, μ_Y, and for which Gaussian fluctuations about the mean of the long-term behavior of the y map were obtained.

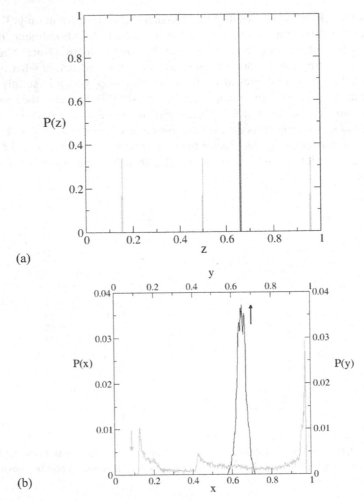

Fig. 4. The behavior (represented as a pdf) of the (a) uncoupled and (b) coupled x_n (blue/grey) and y_n (black) maps at their values of $\mu_X = 3.832$ and $\mu_Y = 2.95$ respectively, for an $\varepsilon = 0.050$.

The effect of the mixing trajectories for one set of parameters is demonstrated in Fig. 4. Part (a) of Fig. 4 shows the long term behavior of the uncoupled ($\varepsilon = 0$) x and y maps for $\mu_X = 3.832$ and $\mu_Y = 2.95$ by plotting the long term behavior of the z map at values of $\mu = \mu_X$ and $\mu = \mu_Y$, respectively, as a probability distribution function (pdf). Once the uncoupled y map (z for $\mu = 2.95$) reaches its equilibrium value (dark/black line), there is a 100% chance of finding it at the fixed point. Likewise, the uncoupled x map (z for $\mu = 3.832$) has reached its period-3 orbit after a few iterations and so there is a 1/3 probability of finding it at any of the three locations (light/blue lines) in its long term behavior when the maps are uncoupled ($\varepsilon = 0$). Part (b) of Fig. 4 demonstrates the pdf behavior in terms of P(x) and P(y) when $\varepsilon = 0.050$. The consequence of mixing for the x map is to effectively push μ_X higher and into the nearby chaotic behavior. While

Table 1. Values of μ_X, μ_Y, and ε

Set	μ_X	μ_Y	ε
1	3.840	3.00	0.025
2	3.830	3.00	0.050
3	3.808	3.00	0.075
4	3.810	2.95	0.025
5	3.832	2.95	0.050
6	3.826	2.90	0.025

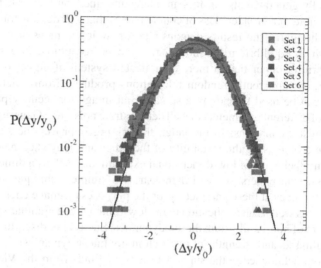

Fig. 5. The normalized values of the long-term behavior of the coupled y_n map relative to the mean as a pdf. The fluctuations from the deterministic map form a nearly Gaussian distribution (solid black line). The symbol legend conforms to the data in Table 1.

the coupling maintains the stability of the y map and pushes the fixed point slightly lower (the mean of y ~ 0.649 for P(y) in Fig. 4(b) is slightly smaller than the fixed point of y = 0.66 in Fig. 4(a)), the coupling introduces fluctuations about the mean even in the long term behavior of the map.

The flatness of these fluctuations about the mean value for P(y) is F = 3 as shown by the box in the plot of Fig. 3, indicating the fluctuations in the tails are nearly Gaussian. While there are examples of large fluctuations elsewhere in the graph, it should be noted that the box highlights an occurrence where the flatness is gently changing and is not discontinuous. To observe this Gaussian more clearly, Fig. 5 is a log plot of $P(\Delta y / y_0)$ where $\Delta y = y - y_{mean}$ and y_0 is the variance of the fluctuations about the mean value. A Gaussian curve is shown for comparison. The fluctuations for all values of μ_X, μ_Y, and ε as listed in Table 1 are plotted. Each histogram in the plot is composed of 20,000 iterations of the coupled y map. A general trend was that for larger values of μ_Y the larger the window in ε that could produce an occurrence of Gaussian statistics for some value of μ_X.

3 Conclusions

We have reported on the results of a pair of coupled logistic maps that produce Gaussian fluctuations about the mean of the stable map due to the unstable map's chaotic behavior. The results give an example of a pair of deterministic equations that together produce a set of equilibrium-like fluctuations in one of the two maps and are applicable to the chaotic encryption of data to hide information within an apparently random, but deterministically generated signal. It is an interesting question for further study if this map is an example of a broader class of coupled maps that will demonstrate this general behavior. Additionally, the results demonstrate a new low dimensional example of a system for which equilibrium thermostatistics can be recaptured in a system driven out of equilibrium, rather than a more complicated system of higher dimensionality. In the former, the apparently random fluctuations produced from a deterministic set of equations can be used to encrypt a signal or an image for security purposes, with multiple sets of different parameters that all demonstrate noise-like Gaussian fluctuations from deterministic equations. In the latter, the mixing improves the stability of one trajectory while increasing the instability of the other and may shed new light on the non-equilibrium behavior of low-dimensional experiments such as a dimer on a shaken plate. In subsequent papers, we will demonstrate a simple technique for using these coupled logistic maps at these parameters for the purposes of image encryption, similar to recent work, as well as an application to the low-dimensional granular system to build a simple picture of non-equilibrium thermodynamics. Such cross-disciplinary results are one of the hallmarks and strengths of research in nonlinear dynamics.

The authors acknowledge the support in part by funds from the Vice Provost for Research and the NSF REU Program at Baylor University.

References

1. Feigenbaum, M.J.: Quantitative universality for a class of nonlinear transformations. J. Stat. Phys. **19**, 25 (1978)
2. Egolf, D.A.: Equilibrim regained: From nonequilibrium chaos to statistical mechanics. Science **287**, 101 (2000)
3. Baxter, G.W., Olafsen, J.S.: Experimental evidence for molecular chaos in granular gases. Phys. Rev. Lett. **99** 028001 (2007)
4. Borges, E.P., Tsallis, C., Ananos, G.F.J., de Oliveira, P.M.C.: Nonequilibrium probabilistic dynamics of the logistic map at the edge of chaos. Phys. Rev. Lett. **89**, 254103 (2002)
5. Pisarchik, A.N., Flores-Carmona, N.J., Carpio-Valadez, M.: Encryption and decryption of images with chaotic map lattices. Chaos **16**, 033118 (2006)
6. Phatak, S.C., Sburesh Rao, S: Logistic map: a possible random number geinerator. Phys. Rev. E, **51**, 3670 (1995)
7. Dorbolo, S., Volfson, D., Tsimring, L., Kudrolli, A.: Dynamics of a bouncing dimer. Phys. Rev. Lett. **95**, 044101 (2005)

8. Luck, J.M., Mehta, A.: Bouncing ball with a finite restitution: Chattering, locking and chaos. Phys. Rev. E **48**, 3988 (1993)
9. Elhadj, Z., Sprott, J.C.: The effect of modulating a parameter in the logistic map. Chaos **18**, 023119 (2008)
10. Pastor-Diaz, I., Lopez-Fraguas, A.: Dynamics of two coupled van der Pol oscillators. Phys. Rev. E **52**, 1480 (1995)
11. Atwell, J., Olafsen, J.S.: Anisotropic dynamics in a shaken granular dimer gas experiment. Phys. Rev. E **71**, 062301 (2005)
12. Oteo, J.A., Ros, J.: Double precision errors in the logistic map: Statistical study and dynamical interpretation. Phys. Rev. E **76**, 036214 (2007)
13. McQuarrie, D.A.: Mathematical Methods for Scientists and Engineers. University Science Books, Sausalito, California (2003)
14. Weiss, J.B.: Moments of the probability distribution for noisy maps. Phys. Rev. A **35**, 879 (1987)
15. Hu, Dong, Y.: Quantum color image encryption based on a novel 3D chaotic system. J. Appli. Phys. **131**, 114402 (2022)

Matching Code Patterns Across Programming Language

Vincent Bushong⊙, Michael Coffey, Austin Lehman, Eric Jaroszewski,
and Tomas Cerny(✉) ⊙

Computer Science, Baylor University, Waco, TX 76798, USA
tomas_cerny@baylor.edu

Abstract. Microservice analysis has seen a surge of interest in recent years
due to the challenges and rewards inherent in analyzing large distributed systems. Microservices are highly decentralized, providing many benefits, including improved performance, shorter development cycles, and enhanced scalability.
However, these benefits come at the cost of hiding knowledge about system operation. Business logic, domain models, and other architectural aspects of microservices are fractured and hidden in the code of individual microservices. To address
this challenge, we developed the Relative Static Structure Analyzer (ReSSA), a
language-agnostic analysis tool driven by small parsers that extract information
from code upon matching user-defined patterns. This paper presents our work in
developing the underlying parsers that power ReSSA definitions for three common
languages used in microservices, NodeJS, Go, and Python. We detail the process
and challenges of parsing the languages into our intermediate format and describe
the benchmark systems we will use as testbeds for our parsers.

Keywords: Code analysis · Parsing · Microservices Abstract Syntax Tree

1 Introduction

Microservice architecture has become the standard approach for new enterprise development of distributed systems. Microservices are an extremely decentralized approach
to software development, resulting in a collection of loosely coupled, independent services that coordinate themselves. Each of these services is independently developed,
deployed, and scaled, resulting in numerous benefits, especially in a distributed cloud
environment.

These benefits come at a cost; since each service is responsible for defining its own
business logic and data model, there is no centralized view of how the system operates at
a high level. Put another way, the system architecture is fragmented and hidden among
the services. To counter this, various methods of Software Architecture Reconstruction (SAR) have been applied to microservices to recover this missing information [4].

This material is based upon work supported by the National Science Foundationunder Grant No.
1854049 and a grant from Red Hat Research.

However, these solutions have been specific to a single language or framework; a single solution that covers multiple platforms has not yet been produced.

To fill this gap, we have produced a method of microservice analysis across platforms, Relative Static Structure Analyzers (ReSSA) [5]. The ReSSA method depends on multiple language parsers to transform code into a Language-Agnostic AST (LAAST), which can then be analyzed by user-defined ReSSAs to extract relevant information. The ReSSA definitions allow users to target specific nodes in the LAAST that contain information pertinent to their specific project.

This paper presents our progress in developing LAAST parsers for three languages commonly used in microservices: NodeJS, Go, and Python. Section 2 details ReSSA and how it can be used to identify relevant portions of microservice definitions. Section 3 describes the development process of individual language parsers and the systems we plan to use as benchmarks for testing the parsers.

2 ReSSA

The ReSSA concept consists of two components: a LAAST and the ReSSA definitions themselves. The goal of the LAAST is to provide a common format for parsing and analysis across languages. This separates the concerns of parsing the individual language and performing pattern matching on the resultant AST. To accomplish this, the LAAST must contain enough relevant information to provide meaningful analysis results while retaining a format relevant to multiple languages. Our goal is to minimize language-specific details that must be addressed within ReSSA definitions.

The LAAST includes classes and their fields and methods, functions, parameters, variable declarations, and various expressions inside function bodies, including call expressions and their arguments and variable usage. Once these elements are parsed into the LAAST, the user defined ReSSAs can perform pattern matching on them to identify important features of, e.g., a microservice, such as endpoint definitions, entity definition, and inter-service calls.

The ReSSA definitions are the second component of the ReSSA ecosystem. These definitions consist of a series of individual pattern matchers for certain node types in the LAAST. End users define the patterns to match using regular expressions, with the ability to extract certain snippets from the code in matched patterns. The extracted information can then be operated on in script callbacks, which perform the logic necessary to transform the raw data into meaningful analysis results.

ReSSA definitions can be nested into tree structures; this allows the user to build a more specific matcher out of the individual node matchers. This feature is natural as the LAAST is also a tree structure that must be traversed and determining whether or not a specific LAAST node is relevant to the user or not often depends on its relationships with other nodes in its subtree.

3 Parser Construction

Python: For parsing through Python, we used the benchmark [1]. This benchmark was mildly complex, as it was structured as a system of general microservices that accessed an array of smaller, more specialized microservices, referred to in the code as "children." These children are what accesses the application's data layer. The purpose of this benchmark is to create a sample microservicebased application in Python other users could use.

Due to the major differences between Python and other object-oriented programming languages, developing a parser for Python proved rather difficult. For one, there is no form of built-in modifiers or annotations in the Python AST that are present in Java AST. Also, while Java requires variable declarations, which the AST could parse, the Python AST does not recognize variable declarations. In order to work around this, we had to figure out what needed to be removed from the Java parser to remove the annotation and modifier functionality without removing any key functionality of the parser. As for the modification with the variable declaration, we modified the parser source code so that it would treat a variable assignment as a declaration, as this would most often be where new variables were introduced or given important new values.

Go: For Go, we used the benchmark [2]. This benchmark consists of 2 services, a movie-microservice, and a user-microservice. The service consists of 21 Go files and 1433 lines of code. This benchmark was designed to show how a microservicebased architecture can be made using Go. In both services, endpoints are called entirely using HTTP methods. The benchmark also utilizes Gin-Gonic, Traefik, and MongoDB, but it is very basic as it serves as an example of a microservice architecture.

Developing the parser proved to be more difficult than initially thought. This is because the Go language's AST was vastly different from any of the others we had seen before. The largest hurdle was that structs were stored in a different subtree of the AST than their member functions. It was even more difficult to determine which struct each member function actually belonged to because that information is stored in a second parameter list.

Our solution was to parse the structs and member functions at different times. First, we began by parsing the structs, leaving their member functions empty, and added them into a HashMap, which mapped the name of each struct to the actual ClassComponent. After parsing all other nodes, we began to parse the member functions of every struct and adding them to a vector. Then we began matching each function's second parameter list to a ClassComponent. After creating a new instance of the ClassComponent and adding the parameter list to it (due to some quirks with Rust), we added it to the vector of components.

NodeJS: To develop a parser for NodeJS, we used the benchmark [3]. This benchmark was not exceedingly complex; however, it proved to be effective in developing the first iteration of the parser. Its simplistic design is effective in achieving the goal the developers intended, which is to introduce microservices in NodeJS. The benchmark comprises 3 services (OrderService, ProductService, and UserService) that span 36 files and consist of 7427 lines of code. The only purpose of these services is to show the architecture of a distributed system in NodeJS. Due to this, the functionality is limited to basic method calls that are used to show the interconnectivity of the services.

In developing the NodeJS parser, we followed an organized path to uncover each individual part of these services. We started by identifying classes described in the benchmark. After these classes were discovered, the methods used in these classes were uncovered. Following the methods, we parsed the bodies of these methods. Finally, the parameters and variables used in the classes and methods were discovered.

The first step we took in developing the NodeJS parser was recognizing classes defined in a service. To find these classes' names, the parser looks for the type "class declaration." It then finds the name of the specific class by looking at the next node in the AST and finding the value with the type "identifier." Once the class name has been discovered, the parser is ready to parse the body of the class. The body of the class is discovered by looking at the sub-tree of the node with the type "class body," specified in the node preceding the declaration of the class. After we have found the class body, we can now investigate the class and discover the methods described in these classes.

To discover the methods and parameters found in these classes, we looked at the sub-trees associated with said classes. The parameters of a class are found by looking for the types "formal parameters" and "parameters" specified in the nodes directly after class declaration and method declaration. The methods defined in these classes are labeled in the AST by the type "method definition," we can then find the name of these methods by looking at the type "property identifier" associated with this node in the AST. The bodies of these classes are labeled by examining the sub-tree of the method and finding the type "statement block," and the remainder of information about the method in question can be found in this sub-tree.

4 Conclusion

We have created parsers for three languages popular for microservice development, NodeJS, Go, and Python. These parsers work with our ReSSA structures to allow users to perform pattern matching for major structural components in microservice systems across languages. We have detailed the challenges associated with each language and our solutions, and we have identified three benchmark systems that will serve as a testbed for future testing.

Our future plans involve crafting ReSSA definitions to extract endpoints, inter-service calls, and entities from each of these benchmark systems to demonstrate the ReSSA's feasibility in performing cross-platform, automated SAR of microservice systems. Another future plan is to develop a case study over a multi-language project in order to show the actual accuracy and applications of ReSSA and the parsers we developed.

Acknowledgement. This material is based upon work supported by the National Science Foundation under Grant No. 1854049 and a grant from Red Hat Research.

References

1. Rajagopalan, S.: pymicro. https://github.com/rshriram/pymicro (2015)
2. Raycad: raycad/go-microservices: Golangmicroservicesexample, https://github.com/raycad/go-microservices
3. Tiwari, S.: Microservices in nodejs. https://github.com/ShankyTiwari/Microservicesin-Nodejs (2019)

4. Walker, A., Laird, I., Cerny, T.: On automatic software architecture reconstruction of microservice applications. In: Kim, H., Kim, K.J., Park, S. (eds.) Information Science and Applications. LNEE, vol. 739, pp. 223–234. Springer, Singapore (2021). https://doi.org/10.1007/978-981-33-6385-4_21
5. Schiewe, M., Curtis, J., Bushong, V., Cerny, T.: Advancing static code analysis with language-agnostic component identification. IEEE Access **10**, 30743–30761 (2022). https://doi.org/10.1109/ACCESS.2022.3160485

Partitionable Programs Using Tyro V2

Arun Sanjel[(✉)] and Greg Speegle

Baylor University, Waco, TX, USA
{arun_sanjel1,greg_speegle}@baylor.edu

Abstract. Extremely large datasets can be efficiently processed by parallel, distributed computation. However, humans are not adept at parallel programming. One solution to this impedance has been the development of APIs which hide the parallelism [6,15,17]. Another is to synthesize parallel programs from sequential ones [2,8,14]. Our approach is to develop a code analysis framework and an execution model which identify the potential components for synthesizing a parallel, distributed program in a high-level API that is equivalent to a given sequential program. Sequential programs which can be translated are called *partitionizable* and the code analysis tool is called *Tyro V2*.

Keywords: Program synthesis · Dataflow programs · Graphs

1 Introduction

Data-Intensive Scalable Computing (DISC) frameworks such as MapReduce [6], Apache Spark [17], and Dask [15] are popular tools for processing large volumes of data. By dividing the input data between many computers, these frameworks gain by both processing the data in parallel and by having a greater percentage of the data fits into memory [5]. However, the underlying architecture of each frameworks is different, potentially requiring substantially greater time and effort to develop these programs compared to sequential programs in languages like Java or Python. As a result, automatically converting a sequential program into a parallel program in the appropriate framework would provide a significant improvement to productivity. In fact, even generating partial solutions can be beneficial.

We previously showed how to use a tool called *Tyro* to convert some sequential Python programs into semantically equivalent PySpark programs [16]. Tyro translates selected code fragments using Abstract Syntax Tree (AST) fragment detection and the GRASSP method [7]. Tyro also checks the generated code using test cases to demonstrate that the input and generated code are equivalent. Tyro is able to extract a loop fragment and convert the loop body to a PySpark processes. However, when the loop body accesses a global variable such as an accumulator, Tyro did not always generate an equivalent program (it failed the test case). This is because Tyro failed to capture the interaction between a loop and variables used outside the loop body.

© The Author(s), under exclusive license to Springer Nature Switzerland AG 2022
H. Han and E. Baker (Eds.): SDSC 2022, CCIS 1725, pp. 185–199, 2022.
https://doi.org/10.1007/978-3-031-23387-6_14

In order to capture such interaction and provide a better translation, we present *Tyro V2*. Tyro V2 uses graph-based loop extraction as a significant improvement for detecting access outside the loop body. We employ graph based flow diagrams to extract the sequential components and transfer them to parallel operations using gradual synthesis, rather than using JSON like *Meta-information* for component detection and extraction. Tyro V2 can also detect the interaction between the components before moving on to the program synthesis step. Early detection not only saves time, but also provides a valid reason why a sequential component cannot be translated using only map operations.

With the improved component detection in Tyro V2, we can now model sequential programs which are excellent candidates for translation. The model identifies *partitionable* programs as ones which can be run in parallel with no modification and *partitionizable* programs which require a reconciliation function. Tyro V2 takes advantage of this concept while attempting to create a DISC application from a sequential program with a reconciliation function. Our approach outlines partitionable and partitionizable programs and how it benefits the program synthesis process.

As such, our paper makes two main contributions:

1. A graph-based tool for program synthesis
2. A computation model as a goal for the synthesis process.

The remainder of the paper is arranged as follows. The second section covers a brief history of our earlier work and DISC application. The new and enhanced Tyro V2 is discussed in Sect. 3. In Sect. 4, we go through our model in terms of DISC applications. Section 5 presents the formal model for partitionizable programs. The paper is concluded in Sect. 6.

2   Related Work

2.1   Tyro

In our first version of Tyro, sequential Python programs are scanned for parallelizable code segments which are converted into equivalent PySpark operations. With the help of an AST, Tyro exacts the program information and stores it in a JSON metadata repository. It then utilizes the Gradual Synthesis for Static Parallelization (GRASSP) [7] approach to convert selected fragments. The key idea behind GRASSP is to gradually grow the translation process i.e. move from simple conversions to complex conversions on subsequent iterations.

Once a program has been converted, the next step is to verify that the PySpark program is equivalent to the original. However, since deciding if two programs are operationally equivalent is undecidable [4], Tyro uses software testing [10] using multiple test cases provided by the user. The verification process is successful if the parallel program passes all of the test cases. If any of the test cases fails, the whole verification fails, and Tyro moves to the next stage of gradual synthesis.

One of the main motivations for graph-based extraction and translation was Tyro's failure of translating interaction between a loop and variables declared outside the loop. Tyro V2 follows the same methodology as its predecessors but with an updated version of the metadata store which includes a dataflow graph. The addition of graph-based extraction and interaction leads to the partitionable model defined in Sect. 5.

2.2 Similar Tools

Casper [2] automatically translates sequential Java code into the MapReduce paradigm. Casper uses two major steps: *Program Synthesis* where it searches for a program summary of each code fragment; and *Code Generation* which generates executable MapReduce code from the program summary. Casper's generated code significantly outperformed the original sequential code.

Tyro shares several aspects with Casper. The initial static code analysis of Casper is similar to the code analysis of Tyro. For example, both tools use ASTs for code analysis. On the other hand, Casper's uses an intermediate representation (IR) of the parsed code for code generations while Tyro V2 uses a graph model of executions. The partitionable model for Tyro V2 is also unique.

One of the first attempts to generate parallel code from sequential programs was MOLD [14]. Similar to Casper, Mold accepts sequential Java programs as input, but MOLD uses a fold operation to identify the candidate loops. The output is a Scala program which can be executed in the Spark environment. The rules for the translation are sound, so the resulting parallel program is equivalent to the original sequential program. However, there are limitations to the complexity of the sequential program. Tyro uses software testing for verification, which allows for more complex programs to be generated.

DIABLO, a process for generating parallel code in [8], also begins with analysis of the AST of a sequential program. Likewise, the framework targets loops as the most likely candidates for parallelization. However, the end result is a monoid comprehension, a computational model which includes standard SQL queries. DIABLO works within a set of restrictions such that for all programs obeying those restrictions, DIABLO generates an equivalent parallel program. Tyro V2 takes a different approach for equivalence (program testing) that allows the possibility of parallelization of programs which do not meet the restrictions. However, Tyro V2 currently does not support programs beyond those covered by DIABLO.

2.3 Graphical Program Representations

Data flow analysis is a method of extracting meaningful data from programs without having to run them. It creates a static summary of information that reflects a program's run-time behavior [12]. Authors in [3] use data flow diagrams similar to an AST to learn different rule-sets for static analyses. Data flow graphs provide a number of advantages over other methods, the most notable of which is their compactness and general direct interpretability [11]. For intermediate representation (IR) of code during program synthesis, the graphical representation is superior to other models for both visual representation and capturing intricate structures.

For example, the Tyro Meta-information was unable to detect interactions between multiple components, while within the graphical representation, each component is a node and edges indicate the interactions. Of course, the more interactions, the more complicated the translation. Thus, the graphical representation identifies both opportunities for parallel code and challenges to the process. Tyro V2 uses the graphical representation described in Sect. 4 to capture these interactions as early as possible.

On the other hand, TensorFlow [1] generates data flow (directed, multi) graphs, which are device-independent intermediate program representations. The Grappler [13], TensorFlow's default graph optimizer, utilizes the created graphs to speed up TensorFlow calculations. The Grappler automatically improves speed by simplifying graphs and performing high-level optimizations that help the majority of target architectures. By improving the mapping of graph nodes to computational resources, such optimization minimizes device peak memory demand and enhances hardware performance. Tyro V2 does not presently leverage graph-based optimization, like TensorFlow does, because the primary objective of such graphical representation in Tyro V2 is to capture component interactions. Enhancing our graph analysis for optimization is ongoing research.

3   Tyro V2

Tyro V2 is built similar to its predecessor. The three major stages for creating a DISC application from a sequential program are shown in Fig. 1. The different stages of Tyro V2 are

1. *Program Analyzer* which accepts the sequential program as input and generates the data flow graph;
2. *Operation Translator* which uses the data flow graph to modify the AST representation of the sequential program;
3. *Code Generator and Verifier* which generates the DISC application and applies the tests.

The rest of this section provides additional details on each stage.

3.1   Program Analyzer

Some information about the input program is required to begin the translation process. Tyro V2 assumes the program is written in python with a restricted set of modules (e.g., numpy is not supported). The Program Analyzer accepts the python program as input and proceeds with three main tasks: parsing source code into an AST, generating a graph of the input program, and static analysis of the program for meta-information. The AST parser for the source code and the static analyzer are unchanged from Tyro [16]. The graph generator is described in detail in Sect. 4.

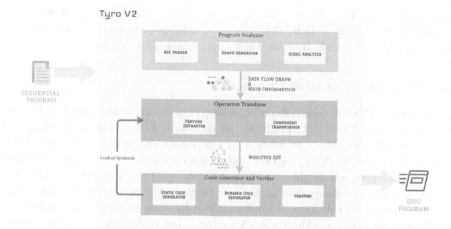

Fig. 1. Tyro V2's system architecture consists of three main components. The first component parses an input Python program into a data flow graph and a JSON description of the operators. The second component translates the operators into the target DISC architecture statement. The third component constructs the DISC program and verifies the results using provided tests. If the verification succeeds, the DISC program is output. If the verification fails, the target DISC architecture is modified and the last two components are executed again.

3.2 Operation Translator

Tyro uses language dependent construction to exchange portions of the original python code with the specific constructs within the target language. For example, PySpark contains a sum() function which can sum values in a dataframe. By analyzing the AST generated in the Program Analyzer, specific transformations can be detected. While the goal of Tyro V2 is to generalize these kind of constructions, the system currently maintains the language dependent operation translator. Future versions of Tyro will replace this component with a graph-based analyzer capable of more general operations.

3.3 Code Generator and Verifier

Tyro generates DISC programs in two steps. The first is a static process which is consistent for all synthesis iterations (e.g., importing the pyspark modules) and the second is the dynamic component which changes as the synthesis process attempts to find more complex solutions. The static process also turns sequential datasets into distributed datasets (e.g., RDDs, Bags). Tyro converts the datasets using the DISC framework's methods (e.g., parallelize() in PySpark). Tyro initially attempts to find a mapper only solution, then proceeds to a map-reduce solution then to alternative attempts (such as implementing join operations).

Tyro V2 begins the verification procedure after each synthesis stage is complete. Tyro V2 runs all of the test cases using the testing suite PyTest. If the program passes the test cases, the resulting executable program returns as an output. If any test case fails, the gradual synthesis proceeds to the next level, repeating the procedure. If all synthesis paths fail, Tyro V2 fails.

3.4 Example Translated Program

Tyro V2 takes Python code containing a loop that iterates over data and converts it into an equivalent DISC program, as shown in Fig. 2. The sequential code passes through all of the previously mentioned steps. Figure 2a is a sequential K-Nearest Neighbors (k-NN) program that uses a user defined function to calculate the distances between neighbors. The *get_neighbors()* method iterates over all of the data points and computes the distance between the point and the input point (x). Tyro identifies this loop as work that can be done in parallel on distributed machines, and generates the PySpark method, *get_neighbors()*. In this method, a SparkContext is created and the training data is converted into an RDD. The *RDD.map()* method is used to apply the user defined function, *dis_cal()*, to every element in the RDD.

4   Graph Based Extraction

Tyro's inability to recognize a global variable or accumulator is the primary motivation for graph-based extraction in Tyro V2. The interactions between a loop fragment and the outside variable, e.g., the accumulator, is a common situation in DISC programs. Tyro V2 utilizes a graph-based extraction method to capture these interactions. The graph-based extraction produces a data flow diagram corresponding to the partitionizable model in Sect. 5. In addition, graph-based extraction is better at visualizing the program's data flow and extracting loops or data-intensive code segments.

Tyro V2 holds a graph G for a program in which a node represents an executable statement. Figure 3b shows an example of a data flow diagram generated by Tyro V2 for the code example in Fig. 3a. Each node is a component that is later grouped together to form the components in the model. An edge between two nodes captures the interaction between these components. Tyro V2 implements three distinct edge lists in the graph: a) program flow, b) loop interaction, and c) global interaction. Figures 3b and 4 depict these three distinct interactions from the three different arrow heads colored: black, green, and red.

The program flow edge-list stores the flow of the given sequential program, represented as black edges in the graph. This program flow edge list form cycles which are potential loop components in the model. The loop interaction are the green arrows in the diagrams. Loop interaction indicates a nested loop structure with dependencies between the inner and outer loops.

Two-loop components interact in Fig. 4, where an inner loop demands more information or data from an outer loop. Because the inner loop is totally dependent on the state of the outer loop, such interaction cannot be removed from the parallel processing. In a distributed context, one possible solution is to introduce

```
# calculate the Euclidean distance between two vectors
def dis_calc(x, y):
    distance = (x[0] - y[0]) ** 2 + (x[1] - y[1]) ** 2
    return sqrt(distance)


# Get distance of points in the dataset
def get_neighbors(train, test_row):
    distances = list()
    for x in enumerate(train):
        dist = dis_calc(test_row, x)
        distances.append((dist, x))
    return distances


# Return Only K nearest neighbours
def KNN(data, query, k):
    data_list = get_neighbors(data, query)
    sortedData = sorted(data_list)
    return sortedData[:k]
```

(a) **Input:** Sequential Python code

```
import pyspark as ps

# calculate the euclidean distance between two vectors
def dis_calc(x, y):
    distance = (x[0] - y[0]) ** 2 + (x[1] - y[1]) ** 2
    return sqrt(distance)


# Get distance of points in the dataset
def get_neighbors(train, test_row):
    distances = list()
    sc = ps.SparkContext()
    train_RDD = sc.parallelize(train)
    distances=train_RDD.map(lambda x:dis_calc(test_row,x),x).collect()
    return distances

# Return only K nearest neighbours
def KNN(data, query, k):
    data_list = get_neighbors(data, query)
    sortedData = sorted(data_list)
    return sortedData[:k]
```

(b) **Output:** Translated PySpark code

Fig. 2. Using Tyro V2 to translate a sequential kNN program into a PySpark implementation. The highlighted section from the input program is translated into the highlighted section in the resulting program. Note that a RDD is created from the list of data, but this could easily be a file for larger data sets. The PySpark program computes the distances between the input point and all other points in parallel. (Color figure online)

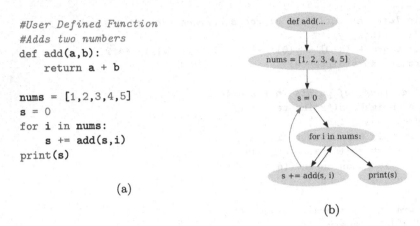

```
#User Defined Function
#Adds two numbers
def add(a,b):
    return a + b

nums = [1,2,3,4,5]
s = 0
for i in nums:
    s += add(s,i)
print(s)
```

(a)

(b)

Fig. 3. A user defined function translated into a data flow graph. The loop is represented by the cycle between the two nodes *for i in nums:* and *s += add(s,i)*. The red line indicates interaction between a node inside of a loop with a node outside of a loop. Within the original Tyro, this could only be handled when the loop body is a user-defined function, as demonstrated here. (Color figure online)

a reconciliation function between two components so that they can operate in parallel. The reconciliation function should combine the resulting output from the inner loops executing in parallel to generate the correct result.

Tyro V2 detects any statement within a loop component that interacts with another statement or component outside the loop using the global interaction edge list. The red arrows in Fig. 4 show global interaction. The two separate loop components interact with the state component *a = 1* in the diagram. Converting all such components is important since the order in which they are executed may affect the final output.

Next, Tyro V2 determines whether or not the interaction can be translated. For simple cases, such as those handled by Tyro, the translation is straightforward. However, for complex interactions, Tyro V2's next step is to generate a reconciliation function that can encapsulate such multi-component interactions. This portion of Tyro V2 is still in progress.

This method of creating a data flow diagram or a graph of a given sequential program not only aids in capturing the interaction of different variables, but it also aids in the extraction of cyclic nodes. Furthermore, with the use of a graph, Tyro V2 can easily separate various components such as state components and loop components that correspond to the concept of partitionable program discussed in Sect. 5.

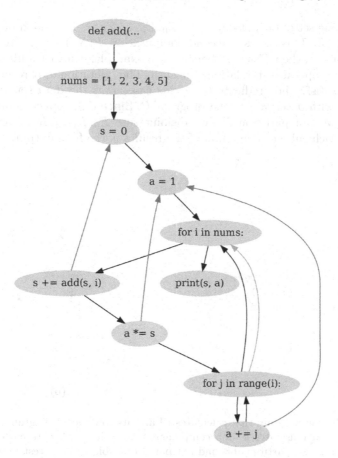

Fig. 4. Tyro models complex interactions between components in a graphical format. The natural flow of the program follows the black edges in the graph. A red edge indicates an interaction between a loop and a variable outside of the loop. The target of the edge is the statement declaring the variable. Whenever a loop interacts with data outside of its scope, simple map processes are not sufficient to represent the computation. A green edge indicates interaction between two loop components – typically a nested loop. Currently, Tyro assumes all nested loops are JOIN operations. If that is incorrect, Tyro fails to find a translation. Handling additional loop interactions are part of our ongoing work. (Color figure online)

5   The Model

Consider a sequential program P over a data set X with parameters Y such that the execution of P on X and Y, denoted $P(X, Y) = Z$, where Z is another data set. Our goal is to partition the input set X into k disjoint sets labeled $X_0, X_1, \ldots X_{k-1}$ such that we can execute a copy of P on each X_i. We represent the partition as $\hat{X}$ and use the notation $P(\hat{X}, Y)$ to represent the execution of P on each element in $\hat{X}$ with parameters Y. The result of this computation is a corresponding set of output data sets, $Z_0, Z_1, \ldots Z_{k-1}$. The union of the output data sets is $\hat{Z}$. This yields a second computation $P(\hat{X}, Y) = \hat{Z}$. If $Z = \hat{Z}$ for all partitions $\hat{X}$, then P is *partitionable*. An example, consider a filter operation such as the σ operator in relational algebra. The operation $\sigma_\theta R$ returns all rows in R that satisfy the predicate θ. In this case, R is the data set and θ is the parameter with σ serving as the program P. Since θ is applied individually to each row, we can partition R into disjoint sets $R_0, R_1, \ldots R_{k-1}$ and perform $\sigma_\theta(R_i)$ on each subset. The union of the results is the final output.

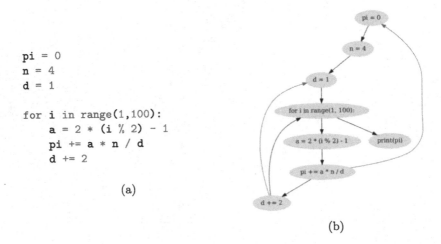

```
pi = 0
n = 4
d = 1

for i in range(1,100):
    a = 2 * (i % 2) - 1
    pi += a * n / d
    d += 2
```

(a)

(b)

Fig. 5. A Python program that calculates PI and its component diagram produced by Tyro V2, which illustrates a loop component interacting with external components. The program is not partitionable and not partitionizable without restructuring.

Unfortunately, while partitionable programs are perfect candidates for large scale parallelism such as DISC, relatively few programs are partitionable. For example, consider the kNN program in Fig. 2a. Partitioning the work between two machines results in two sets of neighbors, not one, which will fail the verification tests.

As a result, we consider the set of programs which can be executed on a partitioned data set and then use a reconciliation function, ρ, which can be applied to $\hat{Z}$ such that $\rho(\hat{Z}) = Z$. We call such programs *partitionizable*. While the ability of partitionizable programs to take advantage of DISC depends somewhat on ρ, being able to execute the initial program in parallel is clearly desirable. For example, the kNN program of Fig. 2a is partitionizable, since the reconciliation function can merge the two sets of neighbors taking the k lowest distances between the sets.

In order to enhance synthesis of the reconciliation function, we break P down into independent components. Figure 6 shows the basic structure. The input component holds the input data (X) while the output component holds the results

(*Z*). The loop component iterates over the data, modifying the state component values as needed. The termination component completes the execution and generates the output.

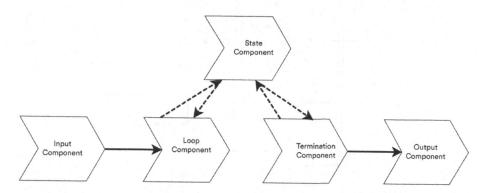

Fig. 6. The sequential computation model for Tyro v2. We assume candidates for DISC translation fit this model. However, we do allow for each of the components themselves to have loop components that can be translated, yielding more complex programs.

As an example, consider the canonical MapReduce program WordCount, which counts the occurrences of each distinct word in a data set. The data set *X* is a text file and there are no parameters *Y*. For sequential execution, the naïve implementation has a state component with an empty dictionary. The loop component reads a line from the data file, breaks the input into words and for each word, sends an update request with the word to the dictionary. On an update, if the word is in the dictionary, the value is increased by one. If the word is not in the dictionary, it is inserted and given a value of 1. The termination component dumps the dictionary to a file.

We can now extend the model to represent partitionizable executions. We call the model in Fig. 7 the distributed model, as it captures the distributed nature of the computation, but does not fully partition the program. First, the input component is divided into *k* units, each of which are read by a distinct loop component. The state component is fundamentally unchanged, except that it atomically accepts requests from the set of loop components. The direction of the dashed lines indicates the data flow. Read requests are represented as dashed arrows going from the state component to the loop component and write requests are dashed arrows in the opposite direction. A new component, the *reconciliation component* is added to the model. The reconciliation component reads the data from the state component, applies *ρ* if needed, and sends the output to the termination component. The termination component is unchanged. For partitionable programs, *ρ* is the identity function. The distributed model is Fig. 7.

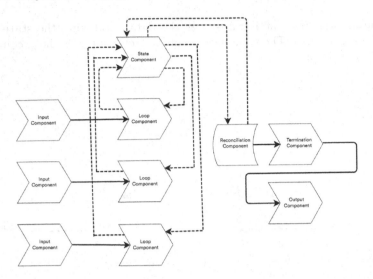

Fig. 7. In the distributed computation model for Tyro v2, the input components are processed in parallel over a distributed set of loop components (identical to the loop components in Fig. 6. Each loop component interacts with a centralized state component. A new Reconciliation Component is added to the sequential model. Note that the reconciliation Component may also be translated to a DISC program.

Under this model, the WordCount program does not change much. The loop components each read a line of data, then atomically send the update request to the state component. No reconciliation is needed, so the termination component dumps this dictionary as before. Clearly, while this model allows some parallel activity, the centralized dictionary is a bottleneck and would need to be removed.

While the distributed model represents distributed computation, the goal for partitionable programs is to eliminate the centralized elements. Thus, the state component should also be distributed with each loop component and each loop component should only communicate with its own local state component. We aggregate these two components into one *execution component*. Now the output of each execution component can be sent to the reconciliation component which generates the final result to be sent to the termination component. This leads to our model for partitionizable computations in Fig. 8.

For program synthesis, the partitionizable model is relatively easy to derive from a sequential program. The termination component is everything after the loop, while the execution program is the remainder of the program. Likewise, it is relatively easy to determine the input to the reconciliation component, as it is the state of the computation after the loop terminates.

Completing our WordCount example, each execution component creates a dictionary over its portion of the input and sends the dictionary to the reconciliation component. In a naïve implementation, all of the dictionaries would go to the same reconciliation component. It would be straightforward for the reconciliation component to merge two dictionaries by applying the update function of one dictionary over the set of entries in the other. A basic aspect of Tyro V2 is to identify and implement merge operations like this.

However, an optimization can also be identified. Since the updates of each dictionary element are independent of the others, we can divide the dictionary into smaller elements and compute the merger in parallel. The smallest point is each entry in each dictionary can be sent to its own ρ function and executed in parallel. Note that this in fact synthesizes the canonical WordCount program for MapReduce. A key future work of Tyro V2 is identifying when optimizations can occur and how to implement them.

Finally, there are programs which are difficult to parallelize even with a reconciliation function. Consider the Pi program in Fig. 5. Clearly, the program is not partitionable since executing multiple instances would generate multiple values for pi, which is incorrect. However, the data to be divided is the number of loop iterations. In this case, dividing the data set into equal sizes yields identical values of pi. Thus, the reconciliation function would have no more information than if the program executed once. A common property of loops in this category is the use of the loop control variable within the computation, which is identified by the Tyro V2 graph.

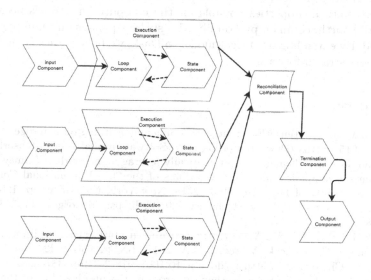

Fig. 8. The most general model for partitionizable computation in Tyro v2 combines the Loop components with their own State Components. Thus, there is no centralized component to handle interactions between the distributed executions. The generated Reconciliation Component must determine the appropriate algorithm to pass the verification tests.

6 Conclusion and Future Work

Efficient and effective program synthesis of a DISC application from a sequential program is still in the early stages of development. In this paper, the original synthesis tool Tyro is extended to include dataflow graphs in order to generate components for a new model of partitionizable executions. The new novel tool Tyro V2 accepts a sequential program and test cases as input and generates a graphical representation that leads to a partitionizable program.

Tyro V2's approach to graphical extraction and representation addressed the issue of interaction between the data processing loop and the remainder of the program. While Tyro recognized and translated specific instances, Tyro V2 models the interaction as a general construct with the goal of synthesizing a reconciliation function to complete the synthesis.

There are several pieces of Tyro V2 that remain to be implemented. First, the synthesis of the reconciliation function is very basic. Significant work on how to extend the current implementation is underway, including examination of machine learning techniques such as programming by example [9]. Second, testing for sequential programs is not always sufficient for testing distributed programs. As a result, synthesis of test cases is also needed. Third, optimization of both the generated code and the synthesis process should be considered. Finally, theoretical properties possible for the computation models need to be addressed. Fourth, common python libraries such as pandas and numpy are not translated. By extending our target model to Dask, we believe these translations will be more straightforward.

References

1. Abadi, M., et al.: TensorFlow: large-scale machine learning on heterogeneous systems (2015). https://www.tensorflow.org/. Software available from tensorflow.org
2. Ahmad, M.B.S., Cheung, A.: Automatically leveraging mapreduce frameworks for data-intensive applications. In: Proceedings of the 2018 International Conference on Management of Data, SIGMOD 2018, New York, NY, USA, pp. 1205–1220. Association for Computing Machinery (2018). https://doi.org/10.1145/3183713.3196891
3. Bielik, P., Raychev, V., Vechev, M.: Learning a static analyzer from data. In: Majumdar, R., Kunčak, V. (eds.) CAV 2017. LNCS, vol. 10426, pp. 233–253. Springer, Cham (2017). https://doi.org/10.1007/978-3-319-63387-9_12
4. Boyer, R.S., Moore, J.S.: A mechanical proof of the unsolvability of the halting problem. J. ACM (JACM) **31**(3), 441–458 (1984)
5. De Oliveira, D.C., Liu, J., Pacitti, E.: Data-intensive workflow management: for clouds and data-intensive and scalable computing environments. Synth. Lect. Data Manag. **14**(4), 1–179 (2019)
6. Dean, J., Ghemawat, S.: MapReduce: simplified data processing on large clusters. Commun. ACM **51**(1), 107–113 (2008)
7. Fedyukovich, G., Ahmad, M.B.S., Bodik, R.: Gradual synthesis for static parallelization of single-pass array-processing programs. SIGPLAN Not. **52**(6), 572–585 (2017). https://doi.org/10.1145/3140587.3062382

8. Fegaras, L., Noor, M.H.: Translation of array-based loops to distributed data-parallel programs. In: Proceedings of the VLDB Endowment, pp. 1248–1260 (2021)
9. Gulwani, S., Pathak, K., Radhakrishna, A., Tiwari, A., Udupa, A.: Quantitative programming by examples. CoRR abs/1909.05964 (2019). http://arxiv.org/abs/1909.05964
10. Jiang, L., Su, Z.: Automatic mining of functionally equivalent code fragments via random testing. In: Proceedings of the Eighteenth International Symposium on Software Testing and Analysis, ISSTA 2009, New York, NY, USA, pp. 81–92. Association for Computing Machinery (2009). https://doi.org/10.1145/1572272.1572283
11. Kavi, K.M., Buckles, B.P., Bhat, U.N.: A formal definition of data flow graph models. IEEE Trans. Comput. **C-35**(11), 940–948 (1986). https://doi.org/10.1109/TC.1986.1676696
12. Khedker, U., Sanyal, A., Karkare, B.: Data Flow Analysis Theory and Practice. CRC Press/Taylor & Francis, Boca Raton (2009)
13. Larsen, R.M., Shpeisman, T.: Tensorflow graph optimizations, March 2022. https://web.stanford.edu/class/cs245/slides/TFGraphOptimizationsStanford.pdf
14. Radoi, C., Fink, S.J., Rabbah, R., Sridharan, M.: Translating imperative code to mapreduce. In: Proceedings of the 2014 ACM International Conference on Object Oriented Programming Systems Languages and Applications, OOPSLA 2014, New York, NY, USA, pp. 909–927. Association for Computing Machinery (2014). https://doi.org/10.1145/2660193.2660228
15. Rocklin, M.: Dask: parallel computation with blocked algorithms and task scheduling. In: Huff, K., Bergstra, J. (eds.) Proceedings of the 14th Python in Science Conference, pp. 130–136 (2015)
16. Sanjel, A.: Tyro: a first step towards automatically generating parallel programs from sequential programs. Ph.D. thesis, Baylor University (2020). https://www.proquest.com/dissertations-theses/tyro-first-step-towards-automatically-generating/docview/2487149689/
17. Shvachko, K., Kuang, H., Radia, S., Chansler, R.: The Hadoop distributed file system. In: 2010 IEEE 26th Symposium on Mass Storage Systems and Technologies (MSST), pp. 1–10. IEEE (2010)

Analyzing Technical Debt by Mapping Production Logs with Source Code

Dipta Das[1] ⓘ, Rofiqul Islam[1], Samuel Kim[1], Tomas Cerny[1](✉) ⓘ, Karel Frajtak[2] ⓘ,
Miroslav Bures[2] ⓘ, and Pavel Tisnovsky[3]

[1] Computer Science, Baylor University, Waco, TX 76798, USA
tomas_cerny@baylor.edu
[2] Computer Science, Czech Technical University, FEE, Prague 12000, Czech Republic
[3] Red Hat, Brno, Czech Republic

Abstract. Poor coding practices, bad design decisions, and expedited software delivery can introduce technical debt. As software grows, manual detection and management of technical debt become increasingly difficult. To address these problems, recent research offers a variety of approaches for automating the process of recognizing and managing technical debt. Unfortunately, current strategies for measuring technical debt depend on static standards that fail to acknowledge software usage patterns in production. In this paper, we utilized existing tools to identify technical debt using static code analysis and then employed production log analysis to rank these debts dynamically.

Keywords: Technical debt · Code smells · Static code analysis · Log analysis

1 Introduction

In today's software business, intense rivalry pushes companies to create their products and deploy new versions under tight time restrictions. Companies frequently use shortcuts in software development and maintenance to achieve these deadlines. These shortcuts, as well as the resulting low software quality, are referred to as technical debt [41]. It employs a financial metaphor to explain the choices between short-term gains and long-term costs in the Software Development Life Cycle (SDLC) [3, 33]. Technical debt can arise as a result of inadequate design decisions and poor coding practices [13, 38]. According to studies, technical debt is largely the outcome of purposeful actions taken to satisfy consumers [41]. It can be a beneficial investment if the project team is aware of its presence and the increased risks it implies [30]. If handled properly, technical debt can help a project achieve its goals sooner or more cheaply [30].

Efficient management of technical debt involves both detection and ranking the issues. Ranking allows the developers to prioritize one task over another while resolving the issues. Several tools can identify technical debt both at the code and design level.

This material is based upon work supported by the National Science Foundation under Grant No. 1854049 and a grant Red Hat Research.

Current approaches of ranking issues use static knowledge such as line of codes, severity, etc. However, software in production shows dynamic behaviors based on the usage pattern. To properly manage technical debt, we need to acknowledge these dynamic behaviors. For example, one part of the code might have a critical issue, but it is merely used in production. In such a scenario, developers should focus on resolving issues that are actively used in production. In this paper, we used existing tools to detect technical debts. Then we utilized both static and dynamic knowledge to properly rank the issues that caused technical debts. Finally, we analyzed the production logs to acknowledge the dynamic usage pattern of the software.

The rest of the paper is organized as follows. Section 2 discusses background and related works. Section 3 describes our proposed method. In Sect. 4, we demonstrate a case study to verify our proposed method. Finally, we conclude the paper in Sect. 5 with a general summary of our contributions along with future works.

2 Background and Related Work

In this section, we describe a general background and related works on technical debt, code smells, architectural degradation, code analysis, and log analysis. In addition, we enumerate through some commonly used tools for detecting technical debt.

2.1 Technical Debt

Technical debt is a future cost attribute that occurs as a result of code smells, architectural flaws, or any other issue in production-level code that needs to be corrected [4, 11, 17, 20–22, 25, 36, 37]. Technical debt can occur for a variety of reasons, the most common of which is the accelerated process of software development [4, 17, 20, 22, 36, 37]. Developers frequently overlook minor needs to complete the project on time and postpone them for future work. It is a common practice in the software development process since clients expect updates in every assessment and a completed product by the deadline. Furthermore, certain accidental and unforeseen defects can only be found when the project is executed on the production platform [22].

Ward Cunningham first introduced the metaphorical term "technical debt" to characterize a specific type of issue that deteriorates software over time [11, 20, 22]. He defined technical debt as follows in his experience report for the OOPSLA conference in 1992:

> "Shipping first time code is like going into debt. A little debt speeds development so long as it is paid back promptly with a rewrite... The danger occurs when the debt is not repaid. Every minute spent on not-quite-right code counts as interest on that debt. Entire engineering organizations can be brought to a stand-still under the debt load of an unconsolidated implementation, object-oriented or otherwise."

Technical debt is regarded in the same way as financial debt [14]. In financial debt, we must refund some extra monetary credit as interest in addition to the capital. Similarly, while attempting to recoup the technical debt, we must spend some extra labor on code, or design refactoring [22].

Martini and Stray [26] identified five common types of technical debt in their work: Code debt, Architectural debt, Test Debt, Documentation debt, and Infrastructure debt. Code debt is primarily caused by various types of code smells. Architectural debt comes as a result of faults in project designs. For example, monolithic architectures are a poor design choice for large projects. Test debt is mostly caused by inadequate test sets and a lack of organized and automated testing. Documentation debt occurs as a result of inadequate levels of code or project documentation, such as insufficient API description, incomplete use-case or domain-model diagram, and so on. Infrastructure debt is mostly the result of poor resource management. Although technical debt arises from many sources, code smells and architectural deterioration are the two most common causes of technical debt.

2.2 Code Smells

Kent Beck introduced the phrase "code-smell" in the 1990s. Code smells, unlike typical bugs, do not always alter the anticipated behavior of the software [10]. Instead, they may have an influence on the performance as well as the future maintainability and scalability. Code smell detection is critical for preventing future defects or difficulties that may arise throughout the SDLC. According to data from a study questionnaire, some of the most common code smells seen by software engineers include duplicated code, large classes, lengthy methods, and so on. These code smells can typically be fixed through some refactoring which not only enhances the code quality but also improves maintainability [1].

2.3 Architectural Degradation

Architectural degradation is the process through which a system's actual architecture deviates from the intended design [6]. This, like code smells, happens when a poor design choice is introduced into a system without consideration for the long-term consequences [40]. Architectural smells such as cyclic dependencies, hublike dependencies, unstable dependencies, cyclic hierarchies, scattered functionality, god components, abstraction without decoupling, multipath hierarchies, ambiguous interfaces, unutilized abstractions, and implicit cross-module dependencies can all indicate architectural deterioration [5].

2.4 Static Code Analysis

Static code analysis [9] creates a representation of the application by recognizing components like classes, methods, fields, and annotations. These representations include Abstract Syntax Trees (AST), Control-Flow Graphs (CFG), or Program Dependency Graphs (PDG). Unlike runtime analysis, such as penetration testing or log analysis, static analysis does not need the deployment of an application, making it more cost-effective. Also, developers can apply static analysis during the application's development phase, which mitigates the risk of inconsistencies in production deployments.

There are two common approaches for static code analysis. Bytecode analysis [2] uses the application's compiled code. In contrast, in source code analysis, we parse through the application's source code without having to compile it into an immediate representation. Source code parsing can be tricky due to different coding conventions; on the other hand, compilers typically normalize bytecode. Besides, bytecode analysis can be utilized as an alternative to source code analysis when we do not have access to the application's source code. Although bytecode analysis can be computationally easier and more accessible, not all languages support bytecode. Bytecode is only available for interpreted languages like JAVA, Python, PHP, etc. This paper uses static code analysis through the SonarQube tool to identify code and design smells along with their severity.

2.5 Tools

There are a plethora of tools and static analyzers available in the market to aid in the detection of code and design smells [15]. These tools differ in many ways based on their input and processing strategies [5]. However, most of them are primarily code-based static analyzers [23]. Some tools consider design or architecture for analysis, while few others analyze the code histories and Git commits.

SonarQube [31, 34], Sonargraph [19], Designite [32], ARCADE [24], Arcan [16], CodeVizard [42], Structure 101 [35] are few examples of code-based tools. SonarQube can operate over 27 programming languages to find code smells like duplicate code, code complexity, security vulnerabilities. In addition, it can detect cyclic dependency architectural smell.

Titan [39], STAN [5], Kaleidoscope [18], Hotspot Detector [8] are design based tools that can detect architectural smells. The prediction subsystem of ARCADE [24] and HistoryMiner [7] are capable of analyzing commit histories to identify code and design smells.

2.6 Log Analysis

Log files, in general, record the state of a system at any given time. These files often provide information on the flow of events inside the system. Log files preserve the necessary knowledge to identify the usage pattern of the software. Furthermore, they give a lot of information on system failures [12]. However, these pieces of information are hard to identify from raw log files. Log parsing and log analysis allow us to extract them from raw log files. But, the formatting of log files is not typically consistent across software, which makes log parsing difficult to generalize [12]. To address this issue, many studies employ a pre-processing step that converts raw logs into some usable format for efficient analysis [27]. The pre-processing might also involve removing redundant or irrelevant data [43].

3 Method

Our proposed method can be divided into four separate modules: technical debt analyzer, function analyzer, log analyzer, and integration module. The architecture of our proposed method is summarized in Fig. 1. The technical debt analyzer module utilizes an external tool such as SonarQube to identify issues through static code analysis. It lists all the issues along with the function names where they reside. It also determines the severity levels of each issue.

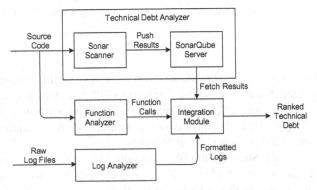

Fig. 1. Architecture of the proposed method

The function analyzer module also takes source code as input and produces a function call graph using static analysis. It first scans all source code files and prepares Abstract Syntax Trees (AST) for each file. Then it traverses through the ASTs using Depth First Search (DFS) and identifies function declarations and function calls. It assigns a unique ID to each function and lists their child node IDs. While traversing the AST, the analyzer also identifies the log statements within each function. Finally, it employs topological sort to find the root node, i.e., the main function.

The log analyzer module takes raw log files as input and processes them to a usable format for further analysis. It scans each log line and figures out the timestamp, log messages, and log types. It filters out all the log lines that do not fall within a given timeframe. Finally, it counts the number of occurrences for each log message. It also maintains the ordering of logs as they appear in the raw log files.

The integration module takes the output from the other three modules and prepares the final output. First, it merges the output of the function analyzer module and log analyzer module by matching the log message and log types. While matching, it also considers the placeholder texts in the log statements. However, there can be conflict while performing this matching. For example, two functions can have identical log messages. In such a case, it considers the sequence of function call graph and the sequence of log messages into account. This merging aims to give each function a weight that signifies how many times it has been called in production.

Next, the integration module merges the output of the technical debt analyzer by mouthing the file and function names. While merging, it sums the severity values for each function node and multiplies them by the occurrence counts of the functions. In other words, it updates the previously assigned weights of each function. Here, the severity values are predefined where higher severity values indicate critical issues. For example, let assume the technical debt analyzer found three issues within a function which severity values are 2, 2, and 3. And the function has been called 10 times in production. The final weight of that function will be $(2 + 2 + 3) * 10 = 70$. The equation for the final weight of a function, therefore, can be written as.

$$w = c \sum_{i=1}^{n} s_i$$

where w is final weight, n is the number of issues, s_i is the severity of an issue, and c is the number of times the function is called in production. Additionally, the severity of an issue s for our purposes was determined through the equation.

$$s = \begin{cases} 1, & \text{if } s_{string} = \text{INFO} \\ 2, & \text{if } s_{string} = \text{Major} \\ 3, & \text{if } s_{string} = \text{CRITICAL} \end{cases}$$

where s_{string} is the string value paired to the "Severity" key within the JSON result produced by SonarQube.

Finally, the integration module sorts each function nodes based on their weights and assigns them a rank. A lower rank signifies a high-priority task. The final output of our proposed method consists of a sorted list of functions along with the issues and their descriptions.

4 Case Study

In our case study, we used cxx-notification-service [29] project from the Red Hat Insights [28]. It is primarily written in Golang. We also collected the raw logs from the production deployment of the service. Our industry partner at Red Hat provides the logs; these are anonymized to hide confidential user data.

First, we prepare the technical debt analyzer module, starting the Sonar- Qube server through a Docker container. We run the docker container in the daemon mode in a separate thread to preserve the analysis results even after our application finishes. It can be also be run separately in a cloud cluster. In that case, we need to provide the URL and authentication credential to connect to the remote SonarQube server. For the remote connection, our application calls the health check API of the SonarQube server to verify it is up and running. The SonarQube server does not perform any analysis. Instead, it works as a storage for analysis results. It also provides a dashboard to visualize the analysis report. Furthermore, it provides intuitive web APIs to fetch and filter those results.

The core technical debt analysis is done by the Sonar scanner application. We run the Sonar scanner application in a Docker container and mount the path of the cxx-notification-service repository. Once the Sonar scanner finishes its analysis, it pushes the results to the SonarQube server. Next, we fetch the result from the SonarQube server using the web API. In our experiment, SonarQube reports 51 issues for the cxx-notification-service repository. Among them, 40 are CRITICAL, 3 are MAJOR, and the remaining 8 are labeled as INFO. A sample issue from those 51 issues is shown in the Listing 1.1. Here, the Function attribute was not provided by the SonarQube, instead we injected it by inspecting the Component and Line attributes. The Function attribute is required for further analysis in the integration module.

Next, the source code is passed to the function analyzer module, which creates a function call graph by analyzing each Golang source file. It also identifies log messages defined within those functions. We analyzed 31 files and identified 190 functions. A sample output from the function analyzer module is shown in Listing 1.2.

The raw log files are inputted into the log analyzer module, which formats the log files by identifying the log message, log level, and occurrence counts within a fixed timeframe. For our experiment, we set the fixed timeframe to twelve hours. Finally, the integration module merges the output of the other three modules and produces the final ranked list as described in our proposed method. Here, we set the severity values for CRITICAL, MAJOR, and INFO level issues to 3, 2, and 1 respectively. A sample block of the final output is shown in Listing 1.3.

Our analysis took a total of 10 min and 45 s to finish. Table 1 depicts the run time distribution among the modules. The technical debt analyzer took the most significant time. However, if we repeat the analysis, the Sonar scanner will use caches and conduct scans just for the modified files. The integration module took the smallest amount of time since it operated only on well-formatted data while other modules had to perform some sort of parsing. The technical debt analyzer, function analyzer, and log analyzer modules are self-contained. So, they can be executed concurrently to minimize total runtime.

4.1 Threats to Validity

In our case study, we analyzed only Golang source code. Modern enterprise applications typically use heterogeneous programming languages that our im plementation fails to acknowledge. However, the architecture of our proposed method is modular, and thus it is possible to integrate separate analyzers for separate languages. Also, we used SonarQube for detecting technical debt that supports all major programming languages. Besides, it is possible to replace SonarQube with any other preferred code analysis tool by utilizing the modular architecture.

Listing 1.1. Sample block of SonarQube result

```
[
  {
    "Component": "ccx:differ/differ.go",
    "Line": 156,
    "Severity": "CRITICAL",
    "Rule": "go:S3776",
    "Type": "CODE_SMELL",
    "Message": "Refactor this method to reduce its Cognitive
    Complexity from 18 to the 15 allowed.",
    "Effort": "8min",
    "Debt": "8min",
    "FilePath": "/ccx-notification-service/differ/differ.go",
    "Function": "processReportsByCluster"
  }
]
```

Listing 1.2. Sample block of function analyzer output

```
[
  {
    "id": 99,
    "name": "setupNotificationProducer",
    "package": "differ",
    "filePath": "/ccx-notification-service/differ/differ.go",
    "logs": [
      {
        "type": "Error",
        "log_msg": "Couldn't initialize Kafka producer with
        the provided config."
      }
    ],
    "childNodeIDs": [
      158,
      22,
      22
    ]
  }
]
```

We used constant values to weigh the severity of the detected debts. However, different companies might prioritize them differently. Allowing users to configure these values will solve the problem.

Listing 1.3. Sample block of final output

```
[
  {
    "Function": "calculateTotalRisk",
    "FilePath": "/ccx-notification-service/differ/differ.go",
    "Rank": 3,
    "Weight": 85,
    "Issues": [
      {
        "Line": 16,
        "Severity": "CRITICAL",
        "Rule": "go:S1192",
        "Type": "CODE_SMELL",
        "Message": "Define a constant instead of duplicating
        this literal types.Timestamp 5 times."
      },
      {
        "Line": 105,
        "Severity": "INFO",
        "Rule": "go:S1135",
        "Type": "CODE_SMELL",
        "Message": "Complete the task associated to this
        TODO comment."
      }
    ]
  }
]
```

Table 1. Run time distribution among the modules

Module	Time (s)
Technical debt analyzer	583
Function analyzer	32
Log analyzer	27
Integration module	3
Total	645

Our log analysis counts the number of times a code portion is called within a fixed timeframe. However, we did not consider user sessions while parsing the logs. Considering a fixed number of user sessions would give us a different perspective in addition to the current fixed timeframe-based analysis.

5 Conclusion and Future Work

Poor coding practices and bad design choices result in technical debt. Automation is required to detect these issues as software grows over time. In addition to detection, ranking of these issues is essential to prioritize tasks for developers.

Although there are a lot of tools to automatically detect code and design faults, they lack the proper knowledge to rank the issues. In this paper, we combined static code analysis with log analysis to detect and rank those issues. Our proposed solution considers software usage patterns which makes the ranking more reliable. In the future, we like to expand our implementation for other languages, including Java and Python. Our current approach takes static log files as input; however, we like to integrate them with real-time log streaming services using the publisher-subscriber model like Apache Kafka. We also plan to experiment with other static analyzer tools to improve the overall runtime and reporting quality. In addition, we are looking for existing and recently published similar work to compare our work to real-world performance. Lastly, we intend for further experimentation to match industry standards through the use of a 24-h period for our log analysis.

Acknowledgement. This material is based upon work supported by the National Science Foundation under Grant No. 1854049 and grant from Red Hat Research.

References

1. Abidi, M., Grichi, M., Khomh, F., Gueheneuc, Y.G.: Code smells for multi-language systems. In: Proceedings of the 24th European Conference on Pattern Languages of Programs. EuroPLop 2019. Association for Computing Machinery, New York (2019). https://doi.org/10.1145/3361149.3361161
2. Albert, E., Gomez-Zamalloa, M., Hubert, L., Puebla, G.: Verification of java byte-code using analysis and transformation of logic programs. In: Hanus, M. (ed.) Practical Aspects of Declarative Languages. pp. 124–139. Springer, Berlin (2007) https://doi.org/10.1007/b98355
3. Amanatidis, T., Mittas, N., Moschou, A., Chatzigeorgiou, A., Ampatzoglou, A., Angelis, L.: Evaluating the agreement among technical debt measurement tools: building an empirical benchmark of technical debt liabilities. Empir. Softw. Eng. **25**(5), 4161–4204 (2020). https://doi.org/10.1007/s10664-020-09869-w
4. Ampatzoglou, A., et al.: The perception of technical debt in the embedded systems domain: An industrial case study. In: 2016 IEEE 8th International Workshop on Managing Technical Debt (MTD), pp. 9–16 (2016). https://doi.org/10.1109/MTD.2016.8
5. Azadi, U., Fontana, F.A., Taibi, D.: Architectural smells de-tected by tools: A catalogue proposal. In: Proceedings of the Scientific Workshop Proceedings of XP2016. XP 2016 Workshops. IEEE Press (2019). https://doi.org/10.1109/TechDebt.2019.00027
6. Baabad, A., Zulzalil, H.B., Hassan, S., Baharom, S.B.: Software architecture degradation in open source software: A systematic literature review. IEEE Access **8**, 173681–173709 (2020). https://doi.org/10.1109/ACCESS.2020.3024671
7. Behnamghader, P., Meemeng, P., Fostiropoulos, I., Huang, D., Srisopha, K., Boehm, B.: A scalable and efficient approach for compiling and an- alyzing commit history. In: Proceedings of the 12th ACM/IEEE International Symposium on Empirical Software Engineering and Mea-surement. ESEM 2018. Association for Computing Machinery, New York (2018). https://doi.org/10.1145/3239235.3239237

8. Biaggi, A., Arcelli Fontana, F., Roveda, R.: An architectural smells detection tool for c and c++ projects. In: 2018 44th Euromicro Conference on Software Engineering and Advanced Applications (SEAA), pp. 417–420 (2018). https://doi.org/10.1109/SEAA.2018.00074

9. Cerny, T., Svacina, J., Das, D., Bushong, V., Bures, M., Tisnovsky, P., Fra- jtak, K., Shin, D., Huang, J.: On code analysis opportunities and challenges for enterprise systems and microservices. IEEE Access pp. 1–22 (2020). https://doi.org/10.1109/ACCESS.2020.301 9985

10. Codegrip: What are code smells? how to detect and remove code smells? (2019). https://www.codegrip.tech/productivity/everything-you-need-to-know-about-code-smells/

11. Cunningham, W.: The WyCash Portfolio Management System. OOPSLA (Mar 1992). http://c2.com/doc/oopsla92.html,

12. Das, D., et al: Failure prediction by utilizing log analysis: A systematic mapping study. In: Proceedings of the International Conference on Research in Adaptive and Convergent Systems, RACS 2020, pp. 188–195. Association for Computing Machinery (2020). https://doi.org/10.1145/3400286.3418263

13. Digkas, G., Ampatzoglou, A., Chatzigeorgiou, A., Avgeriou, P.: On the Temporality of Intro- ducing Code Technical Debt. In: Shepperd, M., Brito e Abreu, F., Rodrigues da Silva, A., P´erez-Castillo, R. (eds.) Quality of Information and Communications Technology, pp. 68– 82. CCIS. Springer International Publishing, Cham (2020). https://doi.org/10.1007/978-3-030-58793-2 6

14. Ernst, N.A., Bellomo, S., Ozkaya, I., Nord, R.L., Gorton, I.: Measure it? manage it? ignore it? software practitioners and technical debt. In: Proceedings of the 2015 10th Joint Meet- ing on Foundations of Software Engineering, ESEC/FSE 2015, pp. 50–60. Association for Computing Machinery, New York (2015). https://doi.org/10.1145/2786805.2786848

15. Falessi, D., Kruchten, P.: Five reasons for including technical debt in the soft- ware engineering curriculum. In: Proceedings of the 2015 European Conference on Software Architecture Workshops, ECSAW 2015. Association for Computing Machinery, New York (2015). https://doi.org/10.1145/2797433.2797462

16. Fontana, F.A., Pigazzini, I., Roveda, R., Tamburri, D., Zanoni, M., Di Nitto, E.: Arcan: A tool for architectural smells detection. In: 2017 IEEE International Conference on Software Architecture Workshops (ICSAW), pp. 282–285 (2017). https://doi.org/10.1109/ICSAW.201 7.16

17. Fontana, F.A., Roveda, R., Zanoni, M.: Technical debt indexes provided by tools: A prelim- inary discussion. In: 2016 IEEE 8th International Workshop on Managing Technical Debt (MTD), pp. 28–31 (2016). https://doi.org/10.1109/MTD.2016.11

18. Haendler, T., Sobernig, S., Strembeck, M.: Towards triaging code-smell can-didates via run- time scenarios and method-call dependencies. In: Proceedings of the XP2017 Scientific Work- shops, XP 2017. Association for Computing Machinery, New York (2017). https://doi.org/10.1145/3120459.3120468

19. Hello2morrow: Sonargraph: Tools to control technical debt and empower softwarecraftsman- ship. https://www.hello2morrow.com/

20. Kruchten, P.: Strategic management of technical debt: Tutorial synopsis. In: 2012 12th Inter- national Conference on Quality Software, pp. 282–284 (2012). https://doi.org/10.1109/QSIC.2012.17

21. Kruchten, P., Nord, R.L., Ozkaya, I.: 4th international workshop on managingtechni- cal debt (mtd 2013). In: Proceedings of the 2013 International Conference on Software Engineering, ICSE 2013. pp. 1535–1536. IEEE Press (2013)

22. Kruchten, P., Nord, R.L., Ozkaya, I., Falessi, D.: Technical debt: Towards a crisper definition report on the 4th international work- shop on managing technical debt. SIGSOFT Softw. Eng. Notes 38(5), 51–54 (2013). https://doi.org/10.1145/2507288.2507326

23. Kruchten, P., Nord, R.L., Ozkaya, I., Visser, J.: Technical debt in soft- ware development: From metaphor to theory report on the third in- ternational workshop on managing technical debt. SIGSOFT Softw. Eng. Notes 37(5), 36–38 (2012). https://doi.org/10.1145/2347696.234 7698

24. Le, D.M., Behnamghader, P., Garcia, J., Link, D., Shahbazian, A., Medvidovic, N.: An empirical study of architectural change in open-source software systems. In: 2015 IEEE/ACM 12th Working Conference on Mining Software Repositories, pp. 235–245 (2015). https://doi.org/ 10.1109/MSR.2015.29

25. Li, Z., Liang, P., Avgeriou, P.: Architectural technical debt identification based on architecture decisions and change scenarios. In: 2015 12th Work- ing IEEE/IFIP Conference on Software Architecture, pp. 65–74 (2015). https://doi.org/10.1109/WICSA.2015.19

26. Martini, A., Stray, V., Moe, N.B.: Technical-, social- and process debt in large-scale agile: An exploratory case-study. In: Hoda, R. (ed.) XP 2019. LNBIP, vol. 364, pp. 112–119. Springer, Cham (2019). https://doi.org/10.1007/978-3-030-30126-2_14

27. Nakka, N., Agrawal, A., Choudhary, A.: Predicting node failure in high perfor-mance computing systems from failure and usage logs. In: 2011 IEEE International Symposium on Parallel and Distributed Processing Workshops and Phd Forum, pp. 1557–1566 (2011)

28. Red Hat: CCX Notification Service (2021). http://github.com/RedHatInsights/ccx-notificat ion-service

29. Red Hat: Red Hat Insights (2021). http://www.redhat.com/en/technologies/management/ins ights

30. Rios, N., Spínola, R.O., Mendonça, M., Seaman, C.: The practitioners' point of view on the concept of technical debt and its causes and consequences: a design for a global family of industrial surveys and its first results from Brazil. Empir. Softw. Eng. 25(5), 3216–3287 (2020). https://doi.org/10.1007/s10664-020-09832-9

31. Saarimaki, N., Baldassarre, M.T., Lenarduzzi, V., Romano, S.: On the accuracy of sonarqube technical debt remediation time. In: 2019 45th Euromicro Conference on Software Engineering and Advanced Applications (SEAA), pp. 317–324 (2019). https://doi.org/10.1109/SEAA. 2019.00055

32. Sharma, T., Mishra, P., Tiwari, R.: Designite - a software design quality assessment tool. In: 2016 IEEE/ACM 1st International Workshop on Bringing Architectural Design Thinking Into Developers' Daily Activities (BRIDGE), pp. 1–4 (2016). https://doi.org/10.1109/Bridge.201 6.009

33. Shull, F., Falessi, D., Seaman, C., Diep, M., Layman, L.: Technical Debt: Showing the Way for Better Transfer of Empirical Results. In: Mu¨nch, J., Schmid, K. (eds.) Perspectives on the Future of Software Engineering: Essays in Honor of Dieter Rombach, pp. 179–190. Springer, Berlin, Heidelberg (2013). https://doi.org/10.1007/978-3-642-37395-4_12

34. SonarQube: Code Quality and Code Security | SonarQube (2021). https://www.sonarqube. org/

35. Structure101: Structural analysis (2021). https://structure101.com/legacy/structural-analysis/

36. Soares de Toledo, S., Martini, A., Przybyszewska, A., Sjøberg, D.I.: Architectural technical debt in microservices: A case study in a large company. In: 2019 IEEE/ACM International Conference on Technical Debt (TechDebt), pp. 78–87 (2019). https://doi.org/10.1109/Tec hDebt.2019.00026

37. Verdecchia, R.: Architectural technical debt identification: Moving forward. In: 2018 IEEE International Conference on Software Architecture Companion (ICSA- C), pp. 43–44 (2018). https://doi.org/10.1109/ICSA-C.2018.00018

38. Verdecchia, R., Kruchten, P., Lago, P.: Architectural technical debt: a grounded theory. In: Jansen, A., Malavolta, I., Muccini, H., Ozkaya, I., Zimmermann, O. (eds.) Software Architecture, pp. 202–219. LNCS. Springer International Publishing, Cham (2020). https://doi.org/ 10.1007/978-3- 030–58923–3 14

212   D. Das et al.

39. Xiao, L., Cai, Y., Kazman, R.: Design rule spaces: A new form of archi- tecture insight. In: Proceedings of the 36th International Conference on Soft- ware Engineering, pp. 967–977. ICSE 2014, Association for Computing Machinery, New York (2014). https://doi.org/10.1145/2568225.2568241

40. Xiao, L., Cai, Y., Kazman, R., Mo, R., Feng, Q.: Identifying and quantify- ing architectural debt. In: Proceedings of the 38th International Conference on Software Engineering, ICSE 2016, pp. 488–498. Association for Computing Machinery, New York (2016). https://doi.org/10.1145/2884781.2884822

41. Yli-Huumo, J., Maglyas, A., Smolander, K.: The sources and approaches to management of technical debt: a case study of two product lines in a middle-size finnish software company. In: Jedlitschka, A., Kuvaja, P., Kuhrmann, M., M¨annisto, T., Mu¨nch, J., Raatikainen, M. (eds.) Product-Focused Software Process Improvement. pp. 93–107. LNCS, Springer International Publishing, Cham (2014). https://doi.org/10.1007/978-3-319-13835-07

42. Zazworka, N., Ackermann, C.: Codevizard: A tool to aid the analysis of software evolution. In: Proceedings of the 2010 ACM-IEEE International Symposium on Empirical Software Engineering and Measurement, ESEM 2010. Association for Computing Machinery, New York (2010). https://doi.org/10.1145/1852786.1852865

43. Zheng, Z., Lan, Z., Park, B.H., Geist, A.: System log pre-processing to improve failure prediction. In: 2009 IEEE/IFIP International Conference on Dependable Systems Networks, pp. 572–577 (2009)

EB-FedAvg: Personalized and Training Efficient Federated Learning with Early-Bird Tickets

Dongdong Li$^{(\boxtimes)}$ (iD)

School of Computer Science and Engineering, South China University of Technology,
Guangzhou 510000, GD, China
201910106942@mail.scut.edu.cn

Abstract. Federated learning is a well-known way to improve privacy in distributed machine learning. Its major goal is to learn a global model that provides good performance to the broadest number of participants. Statistical heterogeneity (also known as non-IID) and training efficiency are two key unresolved concerns in the rapidly developing field of technology. In this paper, we propose Early-Bird FedAvg (EB-FedAvg), a customized federated learning architecture with personalization and training effects based on Early-Bird Tickets. By applying for the early-bird tickets, each client learns an early-bird ticket network (i.e., a sub-network of the base model), and only these early-bird ticket networks are communicated between the server and clients. Instead of learning a shared global model as in traditional federated learning, each client learns a personalized model with EB-FedAvg; communication costs can be greatly reduced due to the compact size of the early-bird ticket network. Experiments on these datasets show that EB-FedAvg outperforms existing systems in personalization, training, and communication cost.

Keywords: Personalization · Efficient federated learning systems · Data heterogeneity

1 Introduction

Federated learning (FL) is a well-liked distributed machine learning framework that enables several clients to train a common global model cooperatively without transmitting their local data [1]. The FL process is coordinated by a central server, and each participating client exchanges only the model parameters with the central server while maintaining local data privacy. By overcoming privacy issues, FL enables machine learning models to learn from decentralized data. FL has been used in numerous real-world situations when data is dispersed across clients and is too delicate to be collected in one location. For instance, FL has been shown to work well when it is used to predict the next word on a smartphone [2].

Participating clients want a shared global model that performs better than their models. Data distribution between clients is non-IID [1, 3]. Due to statistical unpredictability, it's difficult to design a worldwide model that works for all clients. Several research has used FL personalization techniques like meta-learning, multi-task learning, transfer learning, etc. to decrease statistical heterogeneity [4–10].

These solutions frequently entail two steps: 1) developing a global model together, and 2) adapting it for each customer using local data. Two-step customization increases costs. FL's computational cost and diverse devices cause the central server to wait while clients are trained, consuming a lot of energy. Inference speedup and model compression are important FL training advancements. The progressive pruning and training practice involves training a large model, pruning it, and then retraining it to increase performance (the process can be iterated several times). This is a conventional model compression strategy, but recent research links it to more effective training [11]. New research reveals that dense, randomly begun networks contain microscopic subnetworks that, when trained independently, can approach the test accuracy of original networks [12, 13].

Unpredictability in the communication channel between the central server and participating clients might cause transmission delays owing to bandwidth limits. Compressing data between the server and client solves the bottleneck. Sparsification, quantization, etc. [14, 15]. Are common methods. Few efforts have been made to solve both challenges at once. LG-FedAvg may be the only exception [16]. LG-FedAvg was developed with an unreasonable FL configuration, despite each client having enough training data (300 images per class for MNIST and 250 for CIFAR-10).

Our work: We construct EB-FedAvg utilizing Early-Bird Tickets, a bespoke FL framework for training and communication [17]. The Early-Bird Ticket phenomenon helps locate sparse subnetworks within a large base model (EBTNs). Given the same training, EBTNs often outperform a non-sparse base model. Inspired by this fact, we suggest communicating only EBTN parameters between clients and servers in FL after collecting each client's EBTN during each communication round. After adding all of the clients' EBTNs, the server displays the modified EBTN parameters to each client. Finally, each client will learn a tailored model, not a shared global model. The EBTN includes data-dependent features because it's built by trimming the underlying model using local client data. One client's EBTN may not overlap with others when non-IID data is included. After the server completes the aggregate, each EBTN's customization is preserved. Due to the lower EBTN, the required model parameters are also smaller. So, FL's communication efficacy can be boosted.

Our contribution can be summarized as follow:

1. We propose a revolutionary FL framework, namely, EB-FedAvg, that can achieve personalization, more effective training, and effective communication in both IID and non-IID settings;
2. We conduct experiments to compare EB-FedAvg with standalone, FedAvg, Per-FedAvg, and LG-FedAvg [1, 16, 18]. The results of the experiments show that EB-FedAvg is much better than the other methods in terms of personalization, cost of communication, and effectiveness of training.

2 Related Works

Personalization. To achieve personalization, the global model must be modified due to statistical heterogeneity (i.e., the distribution of non-IID data between clients). Personalization is accomplished in existing work by meta-learning, multi-task learning, transfer learning, etc. [4–10]. However, all of the current attempts to achieve personalization through two distinct phases come with additional overhead: 1) A federated global model is learned, and 2) the global model is tailored to each client based on local data.

The Winning Ticket Theory. According to the lottery ticket hypothesis, a fully trained dense network can be pruned to identify a small subnetwork called the winning ticket [12]. By training the isolated winning ticket with the same weight initialization as the dense network's corresponding weights, the dense network can then be trained to have a test accuracy that is comparable to that of the isolated subnetwork. Finding winning tickets requires pricey (iterative) pruning and retraining, though. Morcos et al. investigate the transferability of winning tickets across several datasets [19].

Communication Efficiency. The main barrier for FL is communication because the links between clients and the server frequently function at low rates and can be expensive. Several studies try to lower FL's communication expenses [14, 15]. By combining FedAvg with data compression methods like sparsification, quantization, sketching, etc., the main goal is to reduce the amount of data that is sent between the server and clients.

Efficient Inference and Training. Model compression has been thoroughly investigated for the inference that it is lighter weight. Popular methods include pruning, weight factorization, weight sharing, quantization, dynamic inference, and network architecture search [20–32]. On the other hand, it seems like there is significantly less research on effective training. A small number of studies focus on reducing the total amount of time spent training in situations where people are working in parallel and communicating well [35–37].

3 Design of EB-fedAvg

Early-Bird FedAvg (EB-FedAvg) is an end-to-end combination of Early-Bird Tickets and FedAvg. Figure 1 illustrates an overview of EB-FedAvg. By applying for the Early-Bird Tickets, each participating client discovers an Early-Bird Ticket Network (EBTN). In particular, the EBTN is trained by trimming the base model using the local data of each client. The base model will not be transmitted between the clients and the server, only the parameters of EBTNs. The server will only do the aggregate on the incoming EBTNs after that, and each client will receive the updated parameters for the corresponding EBTNs as a result. After the EBTNs' parameters have been updated, the clients resume their training. Before we explain how to learn local EBTNs and run the global aggregate, we define the following notations that we use in our paper.

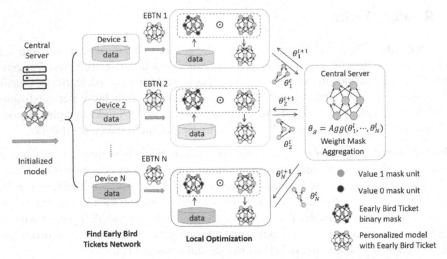

Fig. 1. The EB-FedAvg workflow diagram is displayed. It makes use of Early Bird Tickets to identify the Early Bird Network and, as a result, establish the network structure for Early Learning. Second, sparse weight masks are employed to keep the client personalized during model updates and to increase the effectiveness of data transmission between the client and the server.

Notations: We define $S \subset C$ as a group of clients chosen at random during each training cycle, with $C = \{C_1, \cdots, C_N\}$ representing the N available clients, where C_i signifies the i-th client. S_t denotes the group of clients selected in the t-th round. K denotes the sampling ratio of each round. Let $\theta_i(i \neq g)$ represent the local model parameters on each client C_i, and let θ_g represent the parameters of the base model on the global server. To represent the model parameters of the Early-Bird Ticket Network (EBTN) found in round t, we additionally use the superscript t, θ_i^t. And creates a local early bird ticket binary mask $m_i^t \in \{0, 1\}^{|\theta_i^t|}$. Therefore, the parameters of the relevant weight mask for the client C_i are indicated by the symbol $\theta_i^t \odot m_i^t$. Given the data D_i held by C_i. we split D_i into the training data D_i^{Train}, and test data D_i^{Test}. Pruning probability is p, and the scaling factor for structured pruning is r. The FIFO queue Q has a length of l.

3.1 Training Algorithm

The main distinction between EB-FedAvg and FedAvg is that EBTNs are the only form of communication between clients and the FL server. In each communication round, the EBTNs alone are therefore the only ones used in the server's aggregate. In Algorithm 1, the specifics of the EB-FedAvg training algorithm are described. In general, the training algorithm has the following steps:

Algorithm 1: Training Algorithm of EB-FedAvg

      Data: $(D_1, \cdots, D_N)$ where D_i is the local data on C_i

      Server Executes:

1   initialize the global model θ_g with θ_0

2     $S \leftarrow \{C_1, \cdots, C_N\}$

3   **for** $C_i \in S$ **in parallel do**

4   download θ_g from Global Server

           initialize the client mask m_k^0

5   **end for**

6   **for** each round $t = 1, 2, \ldots$ **do**

7        $k \leftarrow \max(N \times K, 1)$

8        $S_t \leftarrow \{C_1, \cdots C_k\}$

9   **for** each client C_k **in parallel do**

10       $\theta_k^{t+1} \leftarrow ClientUpdate(C_k, \theta_g^t)$

11   **end for**

12     $\theta_g^{t+1} \leftarrow (aggregate\ EBTNs\{\theta_k^{t+1}\})$

13   **end for**

       $ClientUpdate(C_k, \theta_g^t)$:

14     $\theta_k^t = \theta_g^t \odot m_k^t$

15     $\mathcal{B} \leftarrow$ (split local data D_k^{Train} into batches)

16   **for** each local epoch, i from 1 to Epoch **do**

17   **for** batch $b \in \mathcal{B}$ **do**

18           $\theta_k^{t+1} \leftarrow \theta_k^t - \eta \nabla_{\theta_k^t} \ell(\theta_k^t, b)$

19   **end for**

20   Perform structured pruning based on r towards the target ratio p, and generate the mask m_k^t;

21   Calculate the **mask distance** between the current and last subnetworks and add to Q;

22   **if** $Max(Q) < \epsilon$ **then**

23   Updating the early bird ticket binary mask m_k^{t+1}

24   return $\theta_k^{t+1} \odot m_k^{t+1}$ to server

25   **end if**

26   **end for**

27   return $\theta_k^{t+1} \odot m_k^t$ to server

Step I: Send the global server initialization weights θ_g to the client C_i and initialize the client C_i's mask m_k^0.

Step II: The server randomly selects a group of clients S_t given the t-th communication round.

Step III: From the server, each $C_k \in S_t$ client downloads its matching EBTN θ_k^t, where $\theta_k^t = \theta_g^t \odot m_k^t$.

Step IV: Each client C_k start training the local model with θ_k^t. To create a mask m_k^t, structured pruning using the scaling factor r and the pruning probability p. And to store the successively generated subnetworks into a first-in-first-out (FIFO) queue Q with a length of $l = 5$, and calculate the mask m_k^t distances between them.

Step V: Exits when the greatest mask m_k^t distance in the FIFO falls below a predetermined criterion (default 0.1 with normalized distances of $[0,1]$).

3.2 Structured Pruning

Since it is hardware-friendly and best connects to our objective of effective training and performance enhancement, we use the same channel pruning as Liu et al. [21]. To use the scaling factor r in batch normalization (BN) layers as an indicator of channel importance, we follow Liu et al. We generate a mask for each filter separately, and filter channels are pruned under the pruning threshold. To make the pruning threshold easy to use, a percentile of all scaling factors, such as p% of channels, is used to figure it out.

3.3 Early Bird Ticket Masked Weights

The main distinction between EB-FedAvg and existing FL methods is that only the learned weight masks are sent to the central server by each device, which learns an EBTN that only belongs to itself. Figure 2 illustrates how structured pruning can be used to obtain each device's respective EBTN. We can obtain the pruned network mask in response, which we refer to as the Early Bird Ticket Binary Mask (EBTB Mask). To send the mask weights to the central server, we can then use the learned EBTB Mask to perform a real-time structured pruning of the EBTN.

Personalized EBTN   Early Bird Ticket   Mask weights
Binary Mask

Fig. 2. The process of mask weight. Using the EBTN network, which can be acquired after training or updating, we can produce the corresponding early bird ticket binary masks. A mask calculation between the early bird ticket binary mask and the EBTN network can be used to determine the weight mask.

3.4 Aggregate Heterogeneous Weight Masks

The majority of FL methods aggregate data using the FedAvg aggregation strategy, which involves averaging. Instead of updating the full weights in EB-FedAvg, aggregation is done on the weight masks. Additionally, not all elements are overlapped due to the heterogeneity of the weight masks across devices. Because of this, we can't just use an averaging strategy to do aggregation in EB-FedAvg.

Our objective in developing the aggregation strategy is to maximize the retention of personalized data contained in the heterogeneous weight masks. We propose a weighted mask aggregation scheme that accomplishes this goal by independently aggregating each element of the weight mask. As Fig. 3 shows, only elements that appear in two or more weight masks are averaged by the central server. The central server then uses the aggregated values to update these elements in the corresponding weight masks. The central server ignores elements that are not shared in the weight masks and do not aggregate them. Finally, the device will receive the updated weight mask.

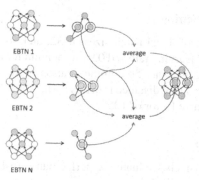

Fig. 3. Weight mask aggregation process. To the server, each device sends its unique weight mask. On the server side, overlapping nodes are aggregated rather than using a strategy that distributes weights evenly as FedAvg does.

3.5 Generate Personalized Model

The weight mask is updated once the weight mask aggregation operation is completed, as shown in Fig. 4. After the weight masks have finished training, the EBTB mask of each device is applied to the updated weight masks one at a time to create a unique model. The EBTB mask makes it more likely that devices will be able to share information while also giving a sparse, personalized model and resolving statistical heterogeneity.

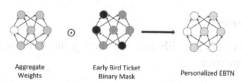

Fig. 4. Personalization via EBTB mask. The client does not update the model immediately after receiving it from the server, unlike FedAvg, in order to preserve the model's personalized nature. The client will choose the received weights using the early bird ticket binary mask to maintain personalization while ensuring information sharing.

4 Evaluation

4.1 Datasets, and Models

In our investigations, we employ the MNIST and CIFAR-10 datasets [38, 40]. We sample the non-IID dataset using the Dirichlet distribution and allocate it to each client with an alpha of 0.2. All of the test sets for the training dataset labels for each client are used to create the evaluation data. We utilized LeNet-5 and AlexNet as our architectures for MNIST and CIFAR-10. In the LeNet-5 and AlexNet designs, we also include a batch-normalization layer after each convolutional layer.

4.2 Hyper-Parameter Setting

We built up 100 clients with local batch sizes of 32, and 50 for local epochs, and an SGD optimizer with a 0.1 learning rate and 0.9 momentum for all experiments. Between the server and the client, there are 50 communication rounds. For structured pruning methods, the threshold for mask distance is 0.1 as well. Additionally, we used a pruning rate of 0.3 and a scale sparse factor of 10^{-4}.

4.3 Compared Methods

Baselines. To comprehensively evaluate the performance of EB-FedAvg, we compare EB-FedAvg against four baselines:

Standalone: each device trains a model independently using only local data without collaborating with other devices [39]. Be aware that using the Standalone technique won't incur any communication costs.

FedAvg is the most classic FL method, and it is employed in commercial products. Devices talk to the central server to send updated local parameters and download the global model so that local training can happen all the time.

Per-FedAvg adds MAML, a prominent meta-learning approach, with FedAvg for customization [18, 41].

LG-FedAvg is a cutting-edge FL method that lets you customize it and improves the efficiency of communication while decreasing the efficiency of computing [16].

4.4 Evaluation Metrics

To evaluate the performance of EB-FedAvg during the training process, we use the following evaluation metrics:

(1) **Inference Accuracy**: We assess the inference accuracy of the test data for each device and report the overall average accuracy for evaluations.

(2) **Communication Cost**: The total number of parameters the model uploads and downloads during training serves as our proxy for the communication cost, which is a significant bottleneck in federal learning.

(3) **Computation Cost**: We quantify the computation time spent on devices for 50 training rounds.

4.5 Training Performance

Inference Accuracy vs. Computation Cost: we compare EB-FedAvg with the baselines in terms of the accuracy-computation tradeoff. Table 1 shows that EB-FedAvg can improve the accuracy of inferences by a lot while reducing the cost of computations by a lot.

First, compared to LG-FedAvg, EB-FedAvg can improve inference accuracy and training computation cost simultaneously. In particular, EB-FedAvg improves inference accuracy by 1.17×, 1.78×, 1.11×, 1.67× on LeNet-5-MNIST, LeNet-5-CIFAR-10, AlexNet-MNIST, and AlexNet-CIFAR-10 in IID, respectively. Inference accuracy improves by 2.96×, 2.23×, 1.79×, and 1.96× on LeNet-5-MNIST, LeNet-5-CIFAR-10, AlexNet-MNIST, and AlexNet-CIFAR-10 in non-IID, respectively.

Second, EB-FedAvg can dramatically reduce computation costs compared to Per-FedAvg, which is specifically designed for personalization. In particular, EB-FedAvg reduces the computation costs of LeNet-5-MNIST, LeNet-5-CIFAR-10, AlexNet-MNIST, and AlexNet-CIFAR-10 in IID by 1.79×, 1.53×, 1.46×, and 1.36×, respectively. In non-IID, EB-FedAvg reduces computation costs by 1.43× and 1.53× on LeNet-5-CIFAR-10 and AlexNet-CIFAR-10, respectively.

Table 1. Comparison between EB-FedAvg and baselines in inference accuracy-computation cost space.

Models	Algorithm	IID		non-IID	
		Acc.(%)	Computation cost. (s)	Acc. (%)	Computation cost. (s)
LeNet-5 MNIST	EB-FedAvg	**96.31**	**673**	**89.56**	780
	FedAvg	96.1	683	89.32	775
	LG- FedAvg	82.31	1281	30.28	703
	Per- FedAvg	94.79	1204	83.87	695
	Standalone	92.14	225	50.14	283
LeNet-5 CIFAR-10	EB-FedAvg	**46.67**	794	**29.31**	839
	FedAvg	46.59	792	27.93	**766**
	LG- FedAvg	26.2	1409	13.06	1410
	Per- FedAvg	36.2	1214	28.2	1200
	Standalone	38.01	353	24.52	372
AlexNet MNIST	EB-FedAvg	**95.09**	**1435**	**88.22**	1369
	FedAvg	94.48	1436	85.57	**1305**
	LG- FedAvg	85.79	2143	49.11	1408
	Per- FedAvg	91.69	2097	83.62	1346
	Standalone	90.29	516	49.89	418
AlexNet CIFAR-10	EB-FedAvg	**49.75**	**1292**	30.09	1377
	FedAvg	48.42	1312	**31.1**	**1360**
	LG- FedAvg	29.73	2007	15.38	2447
	Per- FedAvg	40.77	1762	18.29	2100
	Standalone	36.2	514	24.07	788

Second, EB-FedAvg can dramatically reduce computation costs compared to Per-FedAvg, which is specifically designed for personalization. In particular, EB-FedAvg reduces the computation costs of LeNet-5-MNIST, LeNet-5-CIFAR-10, AlexNet-MNIST, and AlexNet-CIFAR-10 in IID by 1.79×, 1.53×, 1.46×, and 1.36×, respectively. In non-IID, EB-FedAvg reduces computation costs by 1.43x and 1.53x on LeNet-5-CIFAR-10 and AlexNet-CIFAR-10, respectively.

Third, it's not surprising that EB-FedAvg does a lot better than FedAvg in terms of how well it makes inferences and how much it costs to compute. FedAvg is a general FL method that isn't optimized for computation or personalization.

Even though Standalone doesn't have any communication costs, EB-FedAvg does a better job than Standalone because it uses all local data instead of only a few training samples on each device.

Low Communication Cost: Fig. 5 illustrates the comparison of communication costs between EB-FedAvg and the baselines.

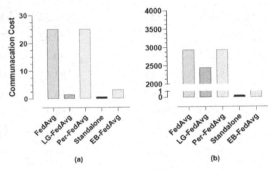

Fig. 5. In terms of communication costs, EB-FedAvg is compared to baselines. LeNet-5 and AlexNet architectures for sending are shown in (a) and (b), respectively. Data sent is measured in megabytes (MB).

As Fig. 5 shows, EB-FedAvg is more communication-efficient in all applications when compared to baselines because of the structured sparsity. Specifically, EB-FedAvg can save 8×, 8.06×, and per-FedAvg on LeNet communication costs, respectively. EB-FedAvg can save 27×, 27.86×, and per-FedAvg on AlexNet communication costs, respectively. EB-FedAvg can save 0.47× and 23.18× in communication costs on LeNet-5-LG-FedAvg and AlexNet-Per-FedAvg, respectively.

The number of Participating Devices: Table 2 shows the BE-FedAvg, which we use to figure out how adding more devices affects how well each communication round works.

We experiment on MNIST and CIFAR10 and change the quantities of participating devices by 5, 20, and 40. As Table 2 illustrates, with more devices participating in each communication round, the inference accuracy marginally improves. For instance, increasing the number of participating devices from 5 to 40 on IID and non-IID enhances the inference accuracy on MNIST-LeNet-5 by 1% and 1.02%, respectively. When the number of devices in IID and non-IID goes from 5 to 40, the inference accuracy on CIFAR-10-LeNet-5 goes up by 1.03% and 1.11%, respectively.

Table 2. The impact of the number of participating devices on EB-FedAvg performance.

Models	Number of devices	IID Acc. (%)	non-IID Acc. (%)
MNIST-LeNet-5	5	96.31	89.56
	20	96.28	90.69
	40	96.49	91.74
CIFAR-10-LeNet-5	5	46.67	29.31
	20	48.78	31.36
	40	47.93	32.57

Data Imbalance Ratio: The amount and type of data on the device have a big effect on how well the FL method works. In practice, there are several bad situations when the data amount is constrained. In addition to having limited data, data on a device frequently displays an imbalance between different classifications. It is challenging for FL strategies to train customized models that perform as well across classes as they do within them. To figure out how the amount of data and the degree of different data types affect the way EB-FedAvg works. We conducted experiments on the MNIST and CIFAR-10 datasets to compare the performance of both datasets on LeNet and AlexNet. The Dirichlet distribution's alpha value, whose greater value roughly equates to a more uniform distribution of data types and amounts among clients, allows us to control the data imbalance rate. The alpha values selected are 0.0001, 0.001, 0.01, and 0.1. As Fig. 6 illustrates, for a fixed number of participating training clients, the accuracy of an inference decreases slightly for smaller alpha values. With the participation of 20 clients,

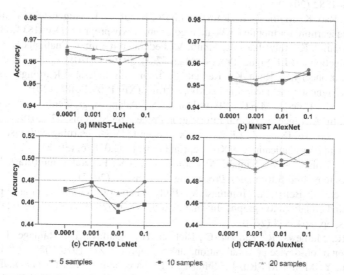

Fig. 6. The impact of the data imbalance rate on EB-FedAvg performance. Three factors are combined in this: data heterogeneity, device heterogeneity, and model heterogeneity. Data heterogeneity refers to data with various balance rates; device heterogeneity refers to the various training devices, and model heterogeneity refers to the utilization of various two models.

the value in Fig. 6(a) dropped from 96.86% to 96.72% when the alpha dropped from 0.1 to 0.0001. In several additional comparisons, the situation is similar. In addition, the increase in the number of clients participating in the training when setting the same alpha value can reduce the impact of the data imbalance degree, in addition to improving the effectiveness of the training. In Fig. 6(a), for example, the accuracy varies between 2% and 4%.

5 Conclusion

We created EB-FedAvg, a customized, effective training and communication FL framework that is motivated by Early-Bird Tickets. The technique removes the generic model parameters from clients' models while maintaining the customized ones by iteratively pruning the neural network channels. Our results show that obtaining the winning ticket at a very early stage, i.e., EB, can achieve the same or better performance than standard training and other personalized architectures. The EBTN obtained through EB-FedAvg not only ensures personalized training but also greatly reduces the consumption during the communication process in the transmission of FL. We assume there are still a lot of promising issues that need to be solved. Testing low-precision EB Train methods on larger models and datasets is an immediate future task. We are also interested in finding out if EB Train could be connected to any further less expensive training methods.

References

1. McMahan, B., Moore, E., Ramage, D., Hampson, S., y Arcas, B.A.: Communication-efficient learning of deep networks from decentralized data. In: Artificial Intelligence and Statistics, pp. 1273–1282 (2017)
2. Hard, A., Rao, K., Mathews, R., Ramaswamy, S., Beaufays, F., Augenstein, S., .Ramage, D.: Federated learning for mobile keyboard prediction. arXiv preprint arXiv:1811.03604. (2018)
3. Li, T., Sahu, A.K., Talwalkar, A., Smith, V.: Federated learning: Challenges, methods, and future directions. IEEE Signal Process. Mag. **37**(3), 50–60 (2020)
4. Jiang, Y., Konečný, J., Rush, K., Kannan, S.: Improving federated learning personalization via model agnostic meta learning. arXiv preprint arXiv:1909.12488. (2019)
5. Khodak, M., Balcan, M.F.F., Talwalkar, A.S.: Adaptive gradient-based meta-learning methods. In: Advances in Neural Information Processing Systems, 32 (2019)
6. Chen, F., Luo, M., Dong, Z., Li, Z., He, X.: Federated meta-learning with fast convergence and efficient communication. arXiv preprint arXiv:1802.07876. (2018)
7. Smith, V., Chiang, C.K., Sanjabi, M., Talwalkar, A.S.: Federated multi-task learning. In: Advances in Neural Information Processing Systems, 30 (2017)
8. Zantedeschi, V., Bellet, A., Tommasi, M.: Fully decentralized joint learning of personalized models and collaboration graphs. In: International Conference on Artificial Intelligence and Statistics, pp. 864–874 (2020)
9. Wang, K., Mathews, R., Kiddon, C., Eichner, H., Beaufays, F., Ramage, D.: Federated evaluation of on-device personalization. arXiv preprint arXiv:1910.10252. (2019)
10. Mansour, Y., Mohri, M., Ro, J., Suresh, A.T.: Three approaches for personalization with applications to federated learning. arXiv preprint arXiv:2002.10619. (2020)
11. Han, S., Mao, H., Dally, W.J.: Deep compression: Compressing deep neural networks with pruning, trained quantization and huffman coding. arXiv preprint arXiv:1510.00149. (2015)

12. Frankle, J., Carbin, M.: The lottery ticket hypothesis: Finding sparse, trainable neural networks. arXiv preprint arXiv:1803.03635. (2018)

13. Liu, Z., Sun, M., Zhou, T., Huang, G., Darrell, T.: Rethinking the value of network pruning. arXiv preprint arXiv:1810.05270. (2018)

14. Konečný, J., McMahan, H.B., Yu, F.X., Richtárik, P., Suresh, A.T., Bacon, D.: Federated learning: Strategies for improving communication efficiency. arXiv preprint arXiv:1610.05492. (2016)

15. Alistarh, D., Grubic, D., Li, J., Tomioka, R., Vojnovic, M.: QSGD: Communication-efficient SGD via gradient quantization and encoding. In: Advances in Neural Information Processing Systems, 30 (2017)

16. Liang, P.P., et al.: Think locally, act globally: Federated learning with local and global representations. arXiv preprint arXiv:2001.01523. (2020)

17. You, H., et al.: Drawing early-bird tickets: Towards more efficient training of deep networks. arXiv preprint arXiv:1909.11957. (2019)

18. Fallah, A., Mokhtari, A., Ozdaglar, A.: Personalized federated learning: A meta-learning approach. arXiv preprint arXiv:2002.07948. (2020)

19. Morcos, A., Yu, H., Paganini, M., Tian, Y.: One ticket to win them all: generalizing lottery ticket initializations across datasets and optimizers. In: Advances in Neural Information Processing Systems, 32 (2019)

20. Li, H., Kadav, A., Durdanovic, I., Samet, H., Graf, H.P.: Pruning filters for efficient convnets. arXiv preprint arXiv:1608.08710. (2016)

21. Liu, Z., Li, J., Shen, Z., Huang, G., Yan, S., Zhang, C.: Learning efficient convolutional networks through network slimming. In: Proceedings of the IEEE International Conference on Computer Vision, pp. 2736–2744 (2017)

22. He, Y., Kang, G., Dong, X., Fu, Y., Yang, Y.: Soft filter pruning for accelerating deep convolutional neural networks. arXiv preprint arXiv:1808.06866. (2018)

23. Wen, W., Wu, C., Wang, Y., Chen, Y., Li, H.: Learning structured sparsity in deep neural networks. In: Advances in Neural Information Processing Systems, 29 (2016)

24. Luo, J. H., Wu, J., Lin, W.: Thinet: A filter level pruning method for deep neural network compression. In: Proceedings of the IEEE International Conference on Computer Vision, pp. 5058–5066 (2017)

25. Liu, S., Lin, Y., Zhou, Z., Nan, K., Liu, H., Du, J.: On-demand deep model compression for mobile devices: A usage-driven model selection framework. In: Proceedings of the 16th Annual International Conference on Mobile Systems, Applications, and Services, pp. 389–400. (2018)

26. Denton, E.L., Zaremba, W., Bruna, J., LeCun, Y., Fergus, R.: Exploiting linear structure within convolutional networks for efficient evaluation. In: Advances in Neural Information Processing Systems, 27 (2014)

27. Wu, J., Wang, Y., Wu, Z., Wang, Z., Veeraraghavan, A., Lin, Y.: Deep k-means: Re-training and parameter sharing with harder cluster assignments for compressing deep convolutions. In: International Conference on Machine Learning, pp. 5363–5372. (2018)

28. Hubara, I., Courbariaux, M., Soudry, D., El-Yaniv, R., Bengio, Y.: Quantized neural networks: training neural networks with low precision weights and activations. J. Mach. Learn. Res. 18(1), 6869–6898 (2017)

29. Wang, Y., Nguyen, T., Zhao, Y., Wang, Z., Lin, Y., Baraniuk, R.: Energynet: Energy-efficient dynamic inference (2018)

30. Wang, Y., et al: Dual dynamic inference: Enabling more efficient, adaptive, and controllable deep inference. IEEE J. Selected Topics Signal Process. 14(4), 623–633 (2020)

31. Shen, J., Wang, Y., Xu, P., Fu, Y., Wang, Z., Lin, Y.: Fractional skipping: Towards finer-grained dynamic cnn inference. In: Proceedings of the AAAI Conference on Artificial Intelligence, vol. 34(4), pp. 5700–5708 (2020)

32. Zoph, B., Le, Q.V.: Neural architecture search with reinforcement learning. arXiv preprint arXiv:1611.01578. (2016)
33. Wang, Y., Xu, C., Xu, C., Xu, C., Tao, D.: Learning versatile filters for efficient convolutional neural networks. In: Advances in Neural Information Processing Systems, 31 (2018)
34. Wang, Y., Xu, C., Xu, C., Tao, D.: Packing convolutional neural networks in the frequency domain. IEEE Trans. Pattern Anal. Mach. Intell. **41**(10), 2495–2510 (2018)
35. Goyal, P., et al.: Accurate, large minibatch sgd: Training imagenet in 1 hour. arXiv preprint arXiv:1706.02677. (2017)
36. Cho, M., Finkler, U., Kumar, S., Kung, D., Saxena, V., Sreedhar, D.: Powerai ddl. arXiv preprint arXiv:1708.02188. (2017)
37. You, Y., Zhang, Z., Hsieh, C.J., Demmel, J., Keutzer, K.: Imagenet training in minutes. In: Proceedings of the 47th International Conference on Parallel Processing, pp. 1–10. (2018)
38. Deng, L.: The mnist database of handwritten digit images for machine learning research [best of the web]. IEEE Signal Process. Mag. **29**(6), 141–142 (2012)
39. Ang, L., Jingwei, S., Xiao, Z., Mi, Z., Hai, L., Yiran, C.: Fedmask: Joint computation and communication-efficient personalized federated learning via heterogeneous masking. In: Proceedings of the 19th ACM Conference on Embedded Networked Sensor Systems (2021)
40. Krizhevsky, A., Hinton, G.: Learning multiple layers of features from tiny images (2009)
41. Finn, C., Abbeel, P., Levine, S.: Model-agnostic meta-learning for fast adaptation of deep networks. In: International Conference on Machine Learning, pp. 1126–1135. (2017)

Author Index

Printed in the United States
by Baker & Taylor Publisher Services